Solutions Manual

to accompany

Air Pollution Control Engineering

Second Edition

AF704365

Noel de Nevers
University of Utah

McGraw-Hill

Boston Burr Ridge, IL Dubuque, IA Madison, WI New York San Francisco St. Louis
Bangkok Bogotá Caracas Lisbon London Madrid
Mexico City Milan New Delhi Seoul Singapore Sydney Taipei Toronto

McGraw-Hill Higher Education

A Division of The ***McGraw-Hill*** *Companies*

Solutions Manual to accompany
AIR POLLUTION CONTROL ENGINEERING

1 2 3 4 5 6 7 8 9 0 TGX/TGX 9 0 9 8 7 6 5 4 3 2 1 0 9

ISBN 0-07-235092-X

http://www.mhhe.com

TABLE OF CONTENTS

Air Pollution Control Engineering, Second Edition
Noel de Nevers
Solutions Manual

This manual contains solutions to all the problems in the text.

Many of those are discussion problems; I have tried to present enough guidance so that the instructor can lead a useful discussion of those problems.

In addition I have added discussion material to many of the computation problems. I regularly assign these as computation, and then after we have agreed that the computation is correct, asked the students what this computation tells them. That leads to discussion. Wherever I can, I begin a discussion of some topic with a computation problem which introduces the students to the magnitudes of various quantities, and thus requires them to read the part of the text covering that topic. Once we know the magnitudes, and have all read that section of the text, we can have an interesting discussion of their meaning.

In this additional discussion I have presented reference when I could. Often I relied on industry "common knowledge", folklore and gossip. I hope I got it all right. If not, I apologize for leading you astray. Where I am not sure about the folklore, I have tried to make that clear in the discussion.

Those problems whose numbers are followed by a double asterisk, ** are ones whose answer is presented in the Answers to Selected Problems in Appendix F of the book.

Most of these problems have been class tested. Some, alas, have not. This manual, as well as the book, is certain to contain errors. I will be grateful to those who point these errors out to me, so that they can be corrected. With the first edition many people brought such errors to my attention. I kept a running correction sheet, and sent copies to anyone who asked for it. If you find such an error, please notify me at

Noel de Nevers
Department of Chemical and Fuels Engineering
50 South Central Campus Drive
University of Utah, Salt Lake City, Utah 84112
Noel.deNevers@m.cc.utah.edu
801-581-6024

Many of the problems go beyond what is in the text, or show derivations which I have left out of the text to make it read easier. I suggest that the instructor tell the students to at least read all the problems, so that they will know what is contained there.

Noel de Nevers
Salt Lake City, Utah, 1999.

Chapter 1

1.1**(a) Applying Bernoulli's equation between the inlet and outlet of the pipe and solving for the pressure difference

$$\Delta P = \frac{\rho V^2}{2}\left(4f\frac{L}{D} + K_{\text{entrance}} + K_{\text{exit}}\right)$$

where K_{exit} is the exit loss coefficient and $K_{entrance}$ is the entrance loss coefficient. From any fluids book we can find that K_{exit} 1.0 and $K_{entrance}$ 0.5. By simple calculations we find that the Reynolds number is $2.3 \cdot 10^9$ and that the relative roughness is 10^{-8}. This puts us far off the typical Moody diagram. So I have extrapolated the smooth tubes curve, estimating $f = 0.0015$. Then $4\,fL/D = 0.173$. This shows that the pipe friction is much less than the entrance and exit losses. The reason is that $L/D = 28.8$ making this a very short pipe. Substituting these values we have

$$\Delta P = \frac{\left(0.075\dfrac{\text{lbm}}{\text{ft}^3}\right)\left(40\dfrac{\text{ft}}{\text{s}}\right)^2}{2}(0.173 + 0.5 + 1)\cdot\frac{\text{lbf s}^2}{32.2\,\text{lbm ft}}\cdot\frac{\text{ft}^2}{144\ \text{in}^2}$$

$$= 0.0216\,\text{psi} \approx 0.02\,\text{psi}$$

(b)

$$\text{Po} = Q\Delta P = 2.63\ \text{E}9\frac{\text{ft}^3}{\text{s}}\cdot 0.0216\frac{\text{lbf}}{\text{in}^2}\cdot\frac{144\ \text{in}^2}{\text{ft}^2}\cdot\frac{\text{HP s}}{550\ \text{ft lbf}}$$

$$= 1.487\ \text{E}7\ \text{HP} = 11.1\ \text{E}6\ \text{kW} \approx 11{,}000\ \text{MW}$$

In the US the total power production capacity divided by the total population is about 2 kw/person, so this equivalent to the power used by 5.5 million people. That is total power including residential use, industrial use, commercial use, street lighting, etc.

The point of this problem is to convince the students that this is not a feasible solution. That does not keep it from being regularly proposed. A variation of it was seriously proposed by high-ranking government officials in Mexico City within the past 10 years; they wanted to blow the polluted air upward out of their mountain basin. See also Prob. 1.5.

1.2 (a). No. That would be true if one ton of CO had the same harmful effects of one ton of SO_2, etc. That is not a safe assumption. Compare the NAAQS (See Chapters 2 and 3) , CO: 9 ppm 8 hour average, SO_2: 0.14 ppm 24 hour average. It appears that a factor of about 10 is involved.

(b) A much harder question. The only really good answer is that there is no “air pollution problem” but rather a family of interrelated problems. It is true that motor vehicles are very substantial contributors.

(c) If possible convince the public there is no single AP problem, but rather a family of problems. If that isn’t possible, then use something like the “Pindex” system devised by EPA which rates pollutant exposures relative to their NAAQSs. I think one averaged or summed those ratios over pollutants, and announced an overall “pollution index”.

1.3 This is discussed in considerable detail in Chapter 12. The principal mechanism of formation of NO (the thermal mechanism) is the reaction $N_2 + O_2 \Leftrightarrow 2\ NO$. This reaction occurs naturally in lightning strokes and in any combustion, e.g. forest fires. But the equilibrium amount of NO is small for temperatures below about 2800 °F. Most combustion occurs at temperatures less than that. In autos and in large-scale industrial furnaces the air is preheated before combustion, raising the final combustion temperature high enough that these process do produce substantial amounts of NO.

1.4 (a) (540/3.1)= 174. So this was 174 times the **annual** man made PM_{10} emissions.
(b) Mixed. Most of that huge amount of material fell within 50 miles of the site, covering a large area up to a few feet thick. However it put enough fine material into the atmosphere that we had world wide reddened sunsets for almost a year.

It is a general proposition that large volcanic eruptions do put enough fine particles (and sulfur oxides which form them in the atmosphere) that they have an effect on world climate for several years. There is a Scientific American article a few years ago about "The Year Without a Summer" (1816) which was caused by a large volcanic eruption.

It is certainly true that, world wide, natural processes put more of the pollutants of interest into the atmosphere than do human processes. This does not mean that what we are doing is unimportant, either locally or globally. Locally, human activities are the major contributor of pollutants in densely populated areas. Globally the amount we emit, e.g. of CO_2 is a small part of the total flux, but enough that it may shift the equilibrium state of the atmosphere.

1.5 This is a variant on the scheme discussed in Example 1.1. It is not likely to work, and was apparently not taken very seriously by the Mexican government. A hovering helicopter causes a strong downward flow of air directly below it. Further away there is a corresponding upward flow. The overall flow is like a horizontal doughnut; with a concentrated flow down in the hole and a much more diffuse flow upward in the surrounding doughnut. This will mix the air in the region, but not remove any pollutants, nor move air out of the region. The hovering helicopters will add their combustion products, which contain some pollutants, to the air stream flowing toward the ground

---

1.6 In economics texts the examples are the application of fertilizer in farming, where x tons per acre produces y incremental yield, but 2x tons per acre produces less than 2y incremental yield, and labor spent weeding the crop, where the first hour has more effect than the second, etc. Students must know that the first hour of studying for a test has more effect on the test score than the second, etc. The situation in air pollution is that when we had no air pollution control equipment, adding any produced a major effect, but now that we have most major sources controlled (in industrial countries) adding more devices will produces less effect per unit of control cost.

Chapter 2

2.1 (a) The volume of the room is 15·15·8 = 1800 ft^3. Assuming that all the sulfur is converted to SO_2, its mass is 5 mg and the concentration is

$$c = \frac{5\text{ E - 3 g}}{1800\text{ ft}^3} \cdot \left(\frac{3.28\text{ ft}}{\text{m}}\right)^3 = 9.8\text{ E - 5}\frac{\text{g}}{\text{m}^3} = 98\frac{\mu\text{g}}{\text{m}^3}$$

(b) This exceeds the annual average NAAQS of 80 μg/m^3.

The sulfur is matches is partly as free sulfur, partly in P_4S_3. Both contribute to the SO_2 formation. Large (2 inch) strike-anywhere matches have about 50 mg of chemicals in their heads. The chemicals are about 5% S, leading to the 2.5 mg of S estimate. For more on match chemistry, see ECT, 20, "Matches".

2.2** (a) $$m = cV = 25\text{ E - 6}\frac{\text{g}}{\text{m}^3} \cdot 10^{-3}\text{m}^3 = 25\text{ E - 9g} = 5.5\text{ E - 11 lb}$$

(b) $$N = \frac{m}{m\text{ of one particle}} = \frac{m}{\rho\frac{\pi}{6}D^3} = \frac{25\text{ E - 9 g}}{\frac{2\text{ g}}{\text{cm}^3} \cdot \frac{\pi}{6} \cdot (1\text{ E - 4 cm})^3} = 2.4\text{ E4}$$

(c) Clearly industry will cite the $0.5 \cdot 10^{-10}$ pounds while environmentalists will cite the 24,000 particles. This shows that two people can both tell the truth, but lead their audience to very different conclusions. We should strive to tell " the whole truth".

2.3 (a) $$m = cV = 80\text{ E - 6}\frac{\text{g}}{\text{m}^3} \cdot 1\text{ E - 3}\frac{\text{m}^3}{\text{breath}} = 80\text{ E - 9}\frac{\text{g}}{\text{breath}}$$

$$= 0.08\frac{\mu\text{g}}{\text{breath}} = 1.8\text{ E - 10}\frac{\text{lb}}{\text{breath}}$$

(b)

$$\text{For SO}_2, M = 64\frac{\text{g}}{\text{mol}}, 1\text{ g} = \frac{\text{mol}}{64} \cdot 6.02\text{ E23}\frac{\text{molecules}}{\text{mol}} = 9.4\text{ E21 molecules}$$

$$N = 80\text{ E9}\frac{\text{g}}{\text{breath}} \cdot 9.4 \cdot 1\text{ E21}\frac{\text{molecules}}{\text{g}} = 7.5\text{ E14}\frac{\text{molecules}}{\text{breath}}$$

See the discussion under Problem 2.2(c). That is applicable here as well.

2.4

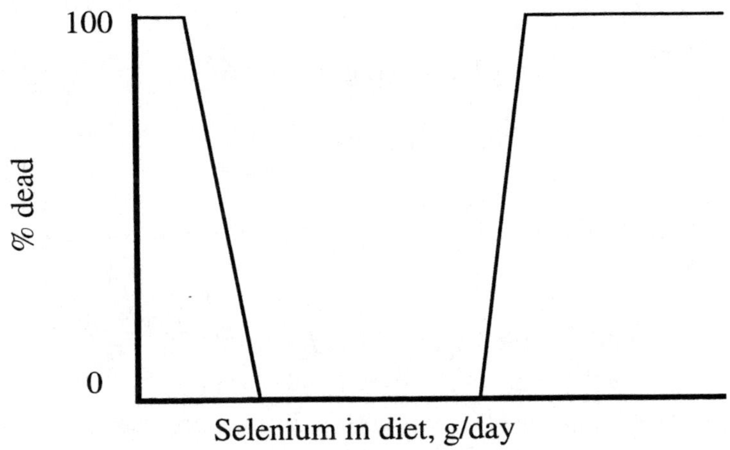

See above plot. This is obviously a **discussion problem.** The same plot is applicable to several metals, e.g. iron, magnesium, zinc. It is also certainly applicable to iodine, which is absolutely necessary, and is added to table salt to avoid goiter, but which was at one time a favorite suicide poison.

We don't really know that the intermediate values are harmless. Maybe that amount of selenium in the diet will kill us in 300 years if we live that long. Currently there are various wildlife refuges where hunters are warned not to eat too much fish or fowl, because of the high selenium content in the water and the bodies of the fish and fowl.

The reason we need some selenium (and other trace metals) is that many of the enzymes in our body, which act as catalysts for important biological reactions, are proteins which must use a metal ion as a template on which to coil into the proper configuration to exert their catalytic activity. Without the right metal ion they are inactive.

2.5 This is speculative but,

(a) Being alone in a hospital bed or nursing home is emotionally more difficult on Christmas and New Years Days than other days, and that must contribute to those death rates.

(b) More elective surgery is scheduled on Monday than later in the week. In elective surgeries for serious problems there is always some risk of a fatality. For example when they had an MD's strike in Israel a few years ago, the daily death rate dropped for a few days, because all elective surgery was postponed.

2.6 This is obviously a fanciful example. The uncontrolled variable is monthly mean temperature, which strongly influences both of the selected variables.

Every epidemiologist hopes to find as clear example as Dr. John Snow did in 1854 when he marked on a map the home of each of the victims of a cholera epidemic, and concluded that they all got their water from the Broad Street Well (or had visited neighbors or eating establishments which did). Closing that well ended the epidemic. Most current epidemiological problems are much more complex than that, and the effects we are looking for are much less dramatic than the pile of bodies after a cholera epidemic. The hazard of overlooking some variable which has a subtle effect of the same magnitude of the effect we are looking for is very real.

2.7 For the assumptions stated

$$\text{Dose} = cQ\Delta t; \quad c = \frac{\text{Dose}}{Q\Delta t}; \log c = \log \text{Dose} - \log Q - \log \Delta t$$

On a plot of log c vs log Δt this is a straight line with slope - 1. For $\Delta t = 0.1$ hr we have

$$c = \frac{10^{-3}\,\text{g}}{1\frac{\text{L}}{\text{hr}} \cdot 0.1\text{hr}} = 0.01\frac{\text{g}}{\text{L}} = 10^{7}\,\frac{\mu\text{g}}{\text{m}^3}$$

Using the conversion shown in the front papers (and $M = 46$ g/mol), we find that $1\text{ppm} = 40.87 \cdot 46 = 1880\frac{\mu\text{g}}{\text{m}^3}$

so that $c = \dfrac{10^7}{1880} = 5319$ ppm. We could use this value to locate where the line of slope -1 crosses the left axis, slightly above the top of the figure.

This problem, which is mostly a discussion problem, shows that the simple assumption that damage is proportional to dose, independent of the rate of administering the dose, is only approximately borne out by the data summarized on this figure. The general shape of the curves (from upper left to lower right) shows that there is some agreement between this simple model and reality. This mediocre agreement is well known from many areas of toxicology. If a person were to receive her lifetime dose of radiation from natural causes all in one minute, that would be swiftly fatal. Clearly the body can deal with more environmental insult, supplied a little at a time than it can deal with applied all at once.

2.8 **This is the logical place for the whole discussion about visibility and its role in US air pollution law. Mostly it has concerned the national parks of the West and Southwest, which historically have had very clean air, and hence very high visual ranges.

(a) $$L_v = \frac{1200\frac{\text{km}\,\mu\text{g}}{\text{m}^3}}{50\frac{\mu\text{g}}{\text{m}^3}} = 24\text{ km} \approx 15\text{ miles}$$

(b) $$c = \frac{1200\frac{\text{kg}\,\mu\text{g}}{\text{m}^3}}{100\text{ mi}} \frac{\text{mi}}{5280\text{ ft}} \cdot \frac{3281\text{ ft}}{\text{km}} = 7.5\frac{\mu\text{g}}{\text{m}^3}$$

(c) $$\text{When } L_v = 20\text{ km},\ c = 60\frac{\mu\text{g}}{\text{m}^3};\ \text{for } c = 61\frac{\mu\text{g}}{\text{m}^3}, L_v = 19.67\text{ km};\ \text{decrease} = 1.63\%$$

$$\text{When } L_v = 200\text{ km},\ c = 6\frac{\mu\text{g}}{\text{m}^3};\ \text{for } c = 7\frac{\mu\text{g}}{\text{m}^3}, L_v = 141.\text{ km};\ \text{decrease} = 14.3\%$$

This shows that a certain increase in particle concentration has more % effect on visual range in a very clean environment than in an average one.

2.9 (a) $\frac{E}{E_o} = 0.02 = \exp(-b_{ext} L_v); \;\; L_v = \frac{\ln 0.02}{b_{ext}} = \frac{3.91}{b_{ext}}$

(b) $$L_v = \frac{3.91}{b_{\text{ext}}} = \frac{1200 \frac{\text{kg}\ \mu\text{g}}{\text{m}^3}}{c}; \;\; b_{\text{ext}} = \frac{3.91\, c}{1200 \frac{\text{kg}\ \mu\text{g}}{\text{m}^3}} = 3.26 \frac{\text{m}^2}{\text{g}} \cdot c$$

(c) $$\text{No. of deciviews} = 10 \ln\left(\frac{b_{\text{ext}}}{1\,\text{E-}5/\text{m}}\right) = 10 \ln\left(\frac{3.91\,\text{E5 m}}{L_v}\right) = 10 \ln\left(\frac{391\ \text{km}}{L_v}\right)$$

This has the advantage that it corresponds to our visual perceptions, which appear to be more logarithmic than linear. The same it true with hearing, which is the reason that sound intensity is expressed in *decibels*. The inventors of the deciview clearly are applying the same idea as the decibel.

(d) If we consider a light path of length L, and give light intensity the symbol E, then

$$E_{\text{in}} - E_{\text{out}} = -E_{\text{in}} \cdot \left(\frac{A_{\text{projected}}}{A}\right) = -E_{\text{in}} \cdot \left(\frac{1.5cL}{D\rho}\right)$$ we now let $L = dx$ and rearrange to

$$\frac{-dE}{E_{\text{in}}} = \left(\frac{1.5c}{D\rho}\right) dx; \;\; \ln \frac{E}{E_{\text{in}}} = \left(\frac{1.5c}{D\rho}\right) \Delta x$$ which is the same as Eq. 2.5 if $b_{\text{ext}} = \left(\frac{1.5c}{D\rho}\right)$

(e) Substituting the values from Example 2.1

$$\frac{b_{\text{ext}}}{c} = \left(\frac{1.5}{D\rho}\right) = \frac{1.5}{(0.3\text{E} - 6\text{m}) \cdot (2\ \text{E6 g.m}^3)} = 2.5 \frac{\text{m}^2}{\text{g}}$$ which is about the middle of the range of values shown in part (b).

2.10 The lines can be satisfactorily represented by

$$\text{Weight Loss}\left(\frac{\text{g}}{100\text{g}}\right) = \frac{0.8}{\text{month}} t + \frac{4.8}{\text{month} \cdot \text{ppm}} t \cdot c$$

Observe that the concentrations are much higher than the NAAQS (0.03 ppm, annual average). Before we replaced coal with natural gas as the principal heating fuel in our cities, the concentrations of particles and SO_2 were much higher than they are now.

2.11 Most sandstones consist of grains of sand, cemented together either by silica or calcium carbonate. Those cemented by calcium carbonate are easily broken down by acid, which dissolves the calcium carbonate. Silica is less soluble in acid, so that a statue made of a sandstone with the sand grains glued together with silica would be expected to be more resistant to acid damage. Limestone or marble statues are destroyed by simple solubility in acid.

Chapter 3

3.1**

$$\frac{\text{emissions}}{\text{consumption}} = \frac{0.41\dfrac{\text{g}}{\text{mile}}}{\dfrac{1\,\text{gal}}{15\,\text{mile}} \cdot \dfrac{6\text{ lb}}{\text{gal}} \cdot \dfrac{454\text{ g}}{\text{lb}}} = 0.00225 = 0.225\%$$

This is about a 90% reduction from pre 1967 cars (see Chapter 13). The purpose of this very simple problem is to make the students look at what an emission standard is.

3.2 ** (a) For the typical coal, (see inside the back cover) ash = 8.7% and heating value = 13,600 Btu/lb, so

$$\frac{\text{permitted emissions}}{\text{uncontrolled emissions}} = \frac{0.03\dfrac{\text{lb}}{10^6\text{ Btu}}}{0.087\dfrac{\text{lb ash}}{\text{lb coal}} \cdot \dfrac{\text{lb coal}}{13{,}600\text{Btu}}} = 0.0047$$

$$\text{Required control efficiency} = 1 - 0.0047 = 0.9953 = 99.53\%$$

(b) This is more than 99%, so the 0.3 lb/10^6 Btu rule is the more restrictive.

(c) For the two rules to be equally restrictive,

$$\frac{\text{permitted}}{\text{controlled}} = 0.01 = \frac{0.03\dfrac{\text{lb}}{10^6\text{ Btu}}}{x\dfrac{\text{lb ash}}{\text{lb coal}} \times \dfrac{\text{lb coal}}{y\text{ Btu}}}$$

For coal with the same heating value as the "typical coal"

$$x = \frac{0.03\dfrac{\text{lb}}{10^6\text{Btu}}}{0.01 \cdot \dfrac{\text{lb coal}}{13{,}600\,\text{Btu}}} = 0.0408 = 4.08\%\,\text{ash}$$

very few coals have ash contents this low, so for most coals the 0.3 lb/10^6 Btu rule is the more restrictive.

(d) When the first NSPS for coal fired power plants was issued (1971) it had only a weight per million Btu value for SO_2. Any coal with less than about 0.5 % S could be burned without scrubbing the exhaust gas, and would meet the standard. The producers of high sulfur coal (and their senators and representatives) pointed out that this would cost their constituents jobs and their campaign contributors profits. So a regulation of this type was written which had the effect of requiring new plants burning low sulfur coal to scrub their exhaust gases to the same percent removal efficiency as plants burning high sulfur coal. One congressional staffer told me (a resident of a low sulfur coal state) that "If we have to have those damn scrubbers, then you have to have them too!" Then by logical extension that kind of regulation was applied to particulates.

3.3

$$\frac{\text{permitted}}{\text{uncotrolled}} = \frac{0.3\dfrac{\text{lb}}{\text{ton}}}{180\dfrac{\text{lb}}{\text{ton}}} = 1.667\ \text{E - 3} = 0.166\%$$

$$\text{controll efficiency} = 1 - 1.667\ \text{E - 3} = 0.99833 = 99.833\%$$

This is more stringent than the regulations for coal fired power plants. The current coal fired power plant emission limit is half that first promulgated in 1971. When those regulations were first promulgated the cement industry cried "Foul!" because they were required to install more efficient particulate control equipment than the coal fired power plant industry was required to install. The coal fired power plant industry replied that their ash was harder to collect, etc. The EPA was simply following the "best technology" wording in the 1970 Clean Air Act, which required that the control for any new source be equivalent to the best which had been demonstrated **in that industry**. They found one or more cement plants which had much more efficient particle control than the best particle control on a coal fired boiler. (Cement plants do have fewer fine particles in their exhaust gases than do coal-fired power plants, so there is some logic in them having to meet a more stringent standard, based on the total weight of particles in their exhaust gases.)

3.4 This is obviously a discussion question. If one takes the "do no harm" viewpoint, then one rejects the plant, and tries to solve the other parts of the country's problems some other way. If one takes the "maximize public welfare" viewpoint, then one accepts the plant, and makes sure to harvest enough revenue from it to make major improvements in the public health, the most important of which is supplying safe drinking water to all the people. Which is right, "do no harm" or "maximize public welfare"? Philosophers have argued about that for at least 3000 years, without finding a clear answer. Engineers have to make this kind of decision, and bear responsibility for the results.

Chapter 4

4.1** $V = 0.2 V_{avg} = 0.2 \cdot \left(39 \frac{ft}{s}\right) = 7.8 \frac{ft}{s}$

$$(cV) = 0.1 (cV)_{avg} = 0.1 \cdot \left(1.2 \frac{lb}{ft^2 hr}\right) = 0.12 \frac{lb}{ft^2 hr}$$

$$c = \frac{cV}{V} = \frac{0.12 \frac{lb}{ft^2 hr}}{7.8 \frac{ft}{s}} \cdot \frac{hr}{3600\ s} = 4.27 \cdot 10^{-6} \frac{lb}{ft^3}$$

The point of this very simple problem is to make the students look at that figure and try to determine what the numbers on it mean. From the values shown it is clear that the concentrations depart much more from being uniform than do the velocities. The reason is almost certainly that centrifugal force carries the largest particles to the outside of the bend, making the concentration lower at the inside.

4.2

$$Q = AV = 48\ ft^2 \cdot 39 \frac{ft}{s} \cdot \frac{60\ s}{minute} = 1872 \frac{ft^3}{s} = 1.123\ E5\ acfm$$

$$\dot{m} = AVc = 1.2 \frac{lb}{ft\ hr} \cdot 48\ ft^2 = 57.6 \frac{lb}{hr}$$

This has the same point as the last problem, to make the students look at that figure.

4.3 Students rarely get this problem right, so it leads to some heated discussions. The goal is to divide the duct into three sub-areas, and sample the middle of each. That is equivalent to dividing it into six sub-areas and placing the sampling point at the boundary between areas (counting from the center) 1 and 2, then 3 and 4 then 5 and 6. Each of the sub-areas should contain 1/6 of the total area. Sketching the areas helps.

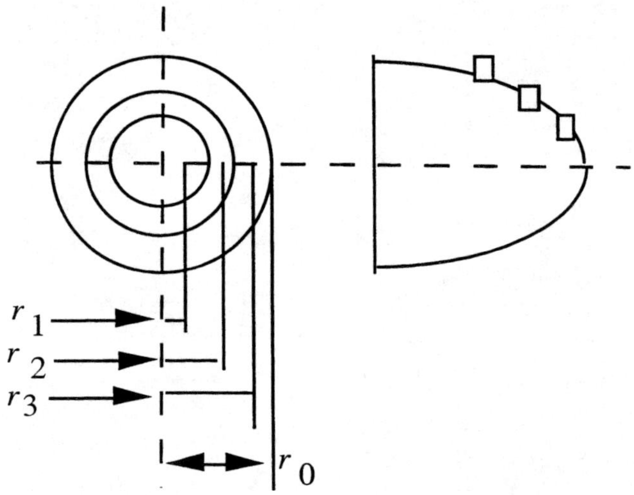

$$\pi r_1^2 = \frac{1}{6} \pi r_0^2; \frac{r_1}{r_0} = \sqrt{\frac{1}{6}} = 0.4082$$

$$\pi r_2^2 = \frac{1}{2} \pi r_0^2; \frac{r_2}{r_0} = \sqrt{\frac{1}{2}} = 0.707$$

$$\pi r_3^2 = \frac{5}{6} \pi r_0^2; \frac{r_3}{r_0} = \sqrt{\frac{5}{6}} = 0.913$$

It also helps in discussing the problem to sketch a laminar (parabolic) velocity profile, and show on it where the three points go, and how they provide a much better estimate of the true average velocity than, for example, a single point in the middle of the duct.

4.4** $$c_{avg} = \frac{\text{total mass}}{\text{total volume}} = \frac{\sum AVc\Delta t}{\sum AV\Delta t} = \frac{\sum Vc}{\sum V}$$
$$= \frac{50 \cdot 0.5 + 60 \cdot 0.3 + 40 \cdot 0.7}{50 + 60 + 40} = 0.4733 \frac{\text{g}}{\text{m}^3}$$

This assumes that velocities and concentrations were constant over each of the 20 minute periods. In practice that is difficult for a source like, for example a basic oxygen steelmaking furnace, whose cycle has rapid changes in particle content.

Many students convert either the velocities or the concentrations to get all in a consistent set of units. The solution shown above makes clear that that is not needed.

4.5 $$\dot{m} = \sum cAV = A\sum cV$$
$$= 1\text{m}^2(10 \cdot 500 + 12 \cdot 600 + 14 \cdot 650 + 15\hat{}675)\frac{\text{m}}{\text{s}} \cdot \frac{\text{mg}}{\text{m}^3} = 31{,}420\frac{\text{mg}}{\text{s}} = 31.42\frac{\text{g}}{\text{s}}$$

This problem simply illustrates how one calculates such total flow rates from the actual measurements.

4.6 Here $y_T = y_{Acid\ Gases} = y_{CO_2} + y_{SO_2}$ and $dy_T = dy_{CO_2} + dy_{SO_2}$ where the $d's$ represent uncertainties. By the stated assumptions, $dy_{CO_2} = 0$, so that

$$\frac{dy_T}{y_T} = \frac{dy_{SO_2}}{y_{CO_2} + y_{SO_2}} = \frac{0.01 \cdot 0.1 \text{ ppm}}{360 \text{ ppm} + 0.1 \text{ ppm}} = \frac{0.001}{360.1} = 2.77\,\text{E} - 6$$

and if our uncertainty in $\frac{dy_{SO_2}}{y_{SO_2}}$ is to be no more than 0.01 =1%, then our uncertainty in the total measured acid gases can be no more than 2.77 E-6. This is a highly impractical requirement, and shows that if there is a large amount of interfering compound in the material to be sampled, one should try very hard to find a different analytical method!

4.7 Assume that all the particles continue straight, so that all the particles which are originally aimed at the probe mouth go into it. Only 0.1 of the gas originally aimed at the probe mouth goes into it, so that the measured concentration will be 10 times the true concentration of the stream being sampled, or 1.0 g/m^3.

A more complex but more satisfactory solution uses Fig 9.18. There we compute that the separation number is 1.23 so that for all three shapes the target efficiency is in the range from 0.6 to 0.8. Thus the estimated concentration in the sample is 0.6 to 0.8 g/m^3.

This problem is discussed in much more detail on pages 575 and 576 of *Sources and Control of Air Pollution* by R. J. Heinsohn and R.L. Kabel, Prentice Hall, Englewood Cliffs, N.J. 1999.

4.8**

$$c_{avg} = \frac{\sum AcV}{\sum V}$$

$$= \frac{8 \cdot 500 + 12 \cdot 600 + 16 \cdot 650 + 20 \cdot 675}{8 + 12 + 16 + 20} = 626.8 \frac{\text{mg}}{\text{m}^3}$$

4.9

$$c_{avg} = \frac{\sum AcV}{\sum V}$$

$$= \frac{(1 \cdot 10 \cdot 500) + (1 \cdot 12 \cdot 600) + (2 \cdot 14 \cdot 650) + (2 \cdot 15 \cdot 675)}{(1 \cdot 10) + (1 \cdot 12) + (2 \cdot 14) + (2 \cdot 15)} = 633.12 \frac{\text{mg}}{\text{m}^3}$$

4.10

$$W_{\text{final}} = 48.07 \cdot (1.001) = 48.118 \text{ g}$$
$$\Delta m = 48.118 - 47.02 = 0.098 \text{ g}$$

$$\text{Reported } c = (\text{correct } c) \cdot \left(\frac{0.098}{0.05}\right) = 24.6 \frac{\mu\text{g}}{\text{m}^3}\left(\frac{0.098}{0.05}\right) = 48.25 \frac{\mu\text{g}}{\text{m}^3}$$

This calculation shows the sensitivity of the results to the proper handling of the sample. It also shows why TSP and PM_{10} results are generally not available until at least a day after the end of the sampling period. Normally the fresh filters are placed in a desiccator for long enough to bring them to equilibrium, then weighed. The same is then done with the used filters. The time for equilibrium is at least a few hours, generally longer.

4.11** (a)

$$m = c\,Q\,\Delta t$$

$$= 25 \frac{\mu\text{g}}{\text{m}^3} \cdot \frac{1\text{E-3m}^3}{\text{min}} \cdot 10 \text{ min} = 0.25 \ \mu\text{g}$$

(b) There are balances which can weigh this small an amount (weigh the difference, not the absolute amount) but they are normally not small, portable and rugged. Most common laboratory scales are useful to about ± 1 mg = 1000 μg.

4.12 (a)

$$20.8 \frac{\mu\text{g}}{\text{m}^3} \cdot \frac{1 \text{ atm}}{0.85 \text{ atm}} = 24.5 \frac{\mu\text{g}}{\text{m}^3 (\text{at standard conditions})}$$

(b) Yes, because to breathe in the same **mass** of air per minutes one would need to breathe in

$$16.7 \frac{\text{L}}{\text{min}} \cdot \frac{1}{0.85} = 19.6 \frac{\text{L}}{\text{min}} \text{ at Salt Lake City.}$$

(c) The correction to sea level over-predicts the mass of particles that one breathes at Salt Lake City, because evidently we actually breathe in the same volume per unit time, not the same mass per unit time. The evidence is that as we go up in elevation, at least as far as 10,000 ft, our body responds by increasing the concentration of red blood cells in our blood, so that we can extract a higher fraction of the oxygen in the air we breathe, and we keep the air flow the same as at sea level. It takes our bodies at least a few days to do this, so that someone who moves rapidly from sea level to a high elevation experiences "shortness of breath" until this adjustment takes place.

4.13 (a) One ton of coal with S% sulfur contains 20 S lb of sulfur. On combustion each pound of sulfur produces 2 pounds of SO_2, so if all the sulfur in the coal formed SO_2 which went to the stack the emission factor would be 40 S lb/ton.

(b) The observed average emission factor (38 S lb/ton) is 95% of that value. All coal ashes contain some alkali or alkaline earth metal oxides. These can collect some SO_2 as sulfates or sulfites. The SO_2 thus collected appears in the ash, not the stack gas. This is the apparent physical explanation of this observation.

4.14 As discussed in Chapter 12, the temperatures in utility and large industrial boilers are much higher than those in commercial and industrial furnaces. Much of the nitrogen oxides are formed from nitrogen and oxygen in the combustion air; the formation is much greater at temperatures higher than about 2800°F than at lower temperatures.

4.15** $$EF_{\text{CO, autos}} = 3.4\frac{g_{CO}}{\text{mile}} \cdot \frac{25\text{mile}}{\text{gal}} \cdot \frac{\text{gal}}{3.785\text{ L}} \cdot \frac{\text{L}}{0.72\text{ kg}} \cdot \frac{\text{kg}}{1000\text{ g}} = 0.0312\frac{g_{CO}}{g_{\text{gasoline}}}$$

$$= 0.0312\frac{g_{CO}}{g_{\text{gasoline}}} \cdot \frac{\text{lb} / 454\text{g}}{\text{ton} / 454 \cdot 2000\text{g}} = 62.4\frac{\text{lb}_{CO}}{\text{ton}_{\text{gasoline}}}$$

Discussion: The point of this problem is to get the students to look at Table 4.2. I regularly take to class with me my copy of AP-42, to show them what it is and how big and detailed it is.

Ask students if either of these is a calculable emission factor, or if both must be based solely on measurements? As far as I can see there is no direct calculation method one can use here so these are simply measured values. Use that as a vehicle for discussing examples 4.4 and 4.5 of the text.

4.16 (a)

$$\text{Emission} = 0.007\frac{\text{ml}}{\text{min}} \cdot \frac{60 \cdot 24\text{ min}}{\text{day}} \cdot \frac{\text{mol}}{22{,}400\text{ ml}} \cdot \frac{28\text{g}}{\text{mol}} = 0.0126\frac{\text{g}}{\text{day}}$$

$$\text{Emission Factor} = \frac{0.0126\ g_{CO} / \text{day}}{600\ g_{\text{food}} / \text{day}} = 2.1\text{ E} - 5\frac{g_{CO}}{g_{\text{food}}}$$

(b) From Problem 4.15 the emission factor for CO from autos is 0.031 g CO/g gasoline, so the ratio of the human respiration CO emission factor to the auto CO emission factor is

$$\text{Ratio} = \frac{2.1\text{E} - 5}{0.031} = 7.0\text{ E} - 4 = \frac{1}{1428}$$

Humans produce much less CO per unit of fuel consumed than autos!

4.17 See Example 4.4.

$$\dot{m}_{\text{parts}} = 1.56\text{ E4 }\frac{\text{lb}}{\text{hr}}$$

$$Q = 179\frac{\text{ton}}{\text{hr}}\cdot 13{,}600\frac{\text{Btu}}{\text{lb}}\cdot 2{,}000\frac{\text{lb}}{\text{ton}} = 4.868\text{ E9 }\frac{\text{Btu}}{\text{hr}}$$

$$\text{Uncontrolled Emissions} = \left(1.56\text{ E4 }\frac{\text{lb}}{\text{hr}}\right) / \left(4.868\text{ E9 }\frac{\text{Btu}}{\text{hr}}\right) = 3.20\frac{\text{lb}}{10^6\text{Btu}}$$

$$\text{Permitted / Uncontrolled} = 0.03 / 3.20 = 9.375\text{ E-3}$$

$$\text{Required Control Efficency} = 1 - 9.375\text{ E-3} = 0.9906 = 99.06\%$$

This is slightly more than 99%, so the 0.03 lb/ million Btu rule is the more restrictive.

4.18 (a) $$\text{EF} = 7\text{A}\frac{\text{lb}}{\text{ton}};\ \text{Ash} = 20\text{A}\frac{\text{lb}}{\text{ton}};\quad \frac{\text{EF}}{\text{Ash}} = \frac{7\text{A}}{20\text{A}} = 0.35 = 35\%$$

(b) $$\text{EF} = 2\text{A},\ \frac{\text{EF}}{\text{Ash}} = \frac{2\text{A}}{20\text{A}} = 0.1 = 10\%$$

To explain the differences requires the students to learn what these devices are. A PC (pulverized coal) furnace grinds the coal to the consistency of face powder, then blows it into the firebox with preheated air. Wall-fired means that the burners are spread out on one wall of the furnace (as opposed to tangentially-fired, in which the burners are at the corners of the furnace, and direct their air-coal mix to create a rotating vortex in the firebox). Wet bottom means that the furnace operates above the melting point of the ash, so that most of the ash forms a liquid on the bottom of the furnace.

A cyclone furnace has a large firebox, with smaller horizontal cylindrical burners on its sides. These operate with gravel-sized coal, hot enough to liquefy the ash, so that most of the ash flows in molten form to the bottom of the furnace. For that reason they produce much less fly ash than PC furnaces.

4.19 (a)

$$\frac{\text{water evaporated}}{\text{ton of coke}} = \frac{\left(2000\frac{\text{lb}}{\text{ton}}\right)\cdot\left(0.325\frac{\text{Btu}}{\text{lb°F}}\right)\cdot(1700 - 212)\text{°F}}{1112\frac{\text{Btu}}{\text{lb}}} = 869\frac{\text{lb water}}{\text{ton coke}}$$

$$\text{particulate emission factor} = 869\frac{\text{lb water}}{\text{ton coke}}\cdot\frac{c}{1\text{ E6}}\frac{\text{lb solids}}{\text{lb water}}$$

$$\text{for } c = 5000\text{ ppm},\ \text{EF} = 4.35\frac{\text{lb}}{\text{ton coke}}$$

(b) This is only fair agreement, because AP-42 shows the same value, independent of c for c 5000 mg/L. They also show a factor of 1.13 lb/ton for water with 1500 mg/L, and values of 1.30 and 0.54 for quenching with the same two kinds of water, with baffles in the quench tower to capture particles. Nonetheless, this shows that this is a "semi-calculable" emission factor, where simple stoichiometry and thermodynamics can give us an answer close to what experimental tests show.

AP-42 also shows the size distribution for all four cases, which we could not have estimated as was done in this example.

4.20 (a)

$$E_{\text{average}} = \frac{(95+98+102)}{3} = 98.33\frac{\text{lb}}{\text{hr}}$$

(b) Using my HP calculator, $s = 3.51$

$$t = \frac{(100 - 98.33)}{s}\sqrt{n} = \frac{(100 - 98.33)\sqrt{3}}{3.51} = 0.824$$

Using the t table in Perry's Chemical Engineers' Handbook, we find that this value of t, for two degrees of freedom corresponds to a probability of about 25%. Thus, based on these three test values, there is a 25% probability that the true average value is greater than 100.

This sort of thing makes lawyers sick to their stomachs. They see the world as black and white, it is true or it is false. The suspect is guilty or innocent. They, and the people who write our air pollution laws hate the probabilistic view of the world. Good engineers know that the probabilistic view is the best view of the world that we have.

Chapter 5

5.1** See page 84. $\frac{t}{D} = \frac{20\,\text{mi}}{16{,}000\,\text{mi}} = 1.25\text{ E-3}$

a) $D = \frac{0.25\,\text{inches}}{1.25\text{ E-3}} = 200\text{ inches} = 16.67\text{ ft}$

b) $t = 1.25\text{ E-3} \cdot 6\text{ in} = 0.0075\text{ inches}$

Discussion: The point of this is to emphasize for them that fact that the atmosphere is thin! It is widely stated that if the earth and a billiard ball were the same diameter, the earth would be smoother. The mountains we all see look pretty rough to us, but. the maximum mountain is perhaps 3 miles (above its base) and earth diameter is about 8000 miles, so 3/8000 = 0.000375. A billiard ball is about 2.5 inches in diameter, so the corresponding maximum roughness would be 0.00037·2.5 inch 0.001 inch. The statement is probably true!

5.2 At the cell boundaries the vertical flow in both cells must be in the same direction (either up or down). With any odd number of cells that condition is satisfied, while with an even number of cells the boundaries must have opposite directed vertical flows at least one of the boundaries. (Sketch the figure and it will be clear!)

5.3 (a) Here if the pitcher throws straight toward the catcher, the ball will not pass over the north pole, because of the eastward velocity of the pitcher. The ball leaving the pitcher's hand has the velocity, directed at the catcher, plus the velocity at right angles due to the pitcher's motion. Both the observer on the space platform and the observer on the earth see the ball pass $4.94 \cdot 10^{-4}$ ft to the side of the North Pole. Both see the ball arrive at the hand of the catcher the same distance to the right as in example 5.1. If the ball were thrown slower, its apparent deviation from a straight line would be greater, as shown by the calculations in Example 5.1.

(b) Here $\sin \varphi$ 0 so the coriolis force plays no role. An observer fixed in space would see the ball move in a straight line with an angle to the equator whose tangent was (the speed of the throw)/(the eastward velocity of the earth). The persons on the earth would see the ball move due North.

(c) Here draw the segment of the circle, showing the position of the pitcher, and the two positions of the catcher, one when the ball is thrown, the other when the ball is caught. Then draw a straight line from the pitcher through the position of the catcher when the ball is thrown. For both directions the position of the catcher when the ball is caught is shifted off this line by the catcher's motion on a circle, while the ball was in flight. In both cases the catcher sees the ball curve to her right.

5.4** (a) The sidereal day should have been used.

(b)We overestimated by a factor of 24/23.93 = 1.00274, or 0.27% (= 1/365)

(c) While the earth makes its trip around the sun we perceive 365.25 days. But someone looking out from the sun or in from the stars would perceive one fewer day, because one of those days is due to our trip around the sun. (If the earth were not rotating on its axis, then we would perceive that stars to remain fixed above us, but would perceive one complete day corresponding to one trip around the sun.)

Data on such astronomical values are in the *Handbook of Chemistry and Physics.*

5.5 Starting with Eq. 5.9 we separate variables and integrate from 1 to 2, finding

$$T_2 = T_1 - \frac{gM\Delta z}{C_P}$$

We then divide by T_1, rearrange and multiply the rightmost term by R/R, to get

$$\frac{T_2}{T_1} = 1 - \frac{R}{C_P} \cdot \frac{gM\Delta z}{RT_1}$$

Then we integrate Eq. 5.8 from 1 to 2, take the exponential of both sides and substitute the value of T_2/T_1 found above to get the final expression of the pressure ratio.

5.6 (a) Taking $T_2 = 0$ and $C_P = 3.5\,R$, we find

$$\Delta z = \frac{3.5RT_1}{gM} = \frac{(3.5)\cdot\left(10.73\frac{\left(\frac{\text{lbf}}{\text{in}^2}\right)\cdot\text{ft}^3}{\text{lb mol °R}}\right)\cdot(519°\,R)}{\left(32.2\frac{\text{ft}}{\text{s}^2}\right)\cdot\left(29\frac{\text{lbm}}{\text{lb mol}}\right)}\cdot\frac{144\,\text{in}^2}{\text{ft}^2}\cdot\frac{32.2\,\text{lbm ft}}{\text{lbf s}^2}$$

$$= 96{,}800 \text{ ft} = 29{,}500 \text{ m}$$

(b) If we had an adiabatic atmosphere, which we don't, then all if it would be below this height, and there would be none above this height.

(c) The equations show that the pressure would be zero. In the "standard atmosphere", at this elevation (29,700 m) the pressure is about 12 mb, indicating that about 1.2% of the atmosphere is above this height.

The point of this problem is to get the students to look at the relations between height, pressure and temperature in the atmosphere.

Why is the atmosphere not adiabatic? Mostly adiabatic applies to quick motions, in which there is little time for heat transfer. Mostly radiation from the ground makes the lower atmosphere non-adiabatic, and interception of solar radiation, mostly by ozone makes the high atmosphere non-adiabatic.

5.7 (a) Start with Eq. 5.7. Deduce from Figure 5.5 that for the standard atmosphere

$$T = 519^\circ R - \left(\frac{0.00356^\circ \mathrm{R}}{\mathrm{ft}}\right) z = a + bz$$

$$\frac{dP}{P} = \frac{-gM}{R}\frac{dz}{a+bz};\ \ln P\Big|_{P_0}^{P} = \frac{-gM}{bR}\ln(a+bz)\Big|_0^z;\ \frac{P}{P_0} = \left(\frac{T}{T_0}\right)^{\frac{-gM}{bR}}$$

Where $b = \dfrac{0.00356^\circ \mathrm{R}}{\mathrm{ft}}$

(b) At the interface

$$\frac{gM}{bR} = \frac{-32.2\dfrac{\mathrm{ft}}{\mathrm{s}^2}\cdot 29\dfrac{\mathrm{lbm}}{\mathrm{lb\,mol}}}{\dfrac{-0.00356^\circ R}{\mathrm{ft}}\cdot 10.73\dfrac{\mathrm{psi\ ft}^3}{\mathrm{lb\,mol\ {}^\circ R}}\cdot\dfrac{144\,\mathrm{in}^2}{\mathrm{ft}^2}\cdot 32.2\dfrac{\mathrm{lbm\ ft}}{\mathrm{lbf\ s}^2}} = -5.27$$

$$\frac{P}{P_0} = \left(\frac{390^\circ R}{519^\circ R}\right)^{(-(-5.27))} = 0.222$$

(c) For constant T (in the stratosphere) we integrate Eq 5.7 to

$$P = P_{\mathrm{interface}}\exp\left\{\frac{-Mg(z - z_{\mathrm{interface}})}{R\,T_{\mathrm{stratosphere}}}\right\}$$

5.8 Below the mixing height $T = 20^\circ\mathrm{C} - 9.7\dfrac{^\circ\mathrm{C}}{\mathrm{km}}z$, while above the mixing height $T = 15^\circ\mathrm{C} - 6.5\dfrac{^\circ\mathrm{C}}{\mathrm{km}}z$. The mixing height is the elevation at which these two are equal. Eliminating T between them and solving for z, we find $z = 1.56$ km = 1560 m = 5162 ft. This is approximately the value shown on Fig. 5.7.

5.9 For this condition the average molecular weight is

$$M_{\mathrm{avg}} = 0.9885(29) + 0.0115(18) = 28.87\frac{\mathrm{g}}{\mathrm{mol}}$$

and the average heat capacity is

$$C_{P_{\mathrm{avg}}} = 0.9885(3.5\,R) + 0.0115(4.1\,R) = 3.5069\,R$$

So that $\dfrac{M}{C_P} = \dfrac{28.87}{3.5069} = \dfrac{8.232\dfrac{\mathrm{g}}{\mathrm{mol}}}{R}$. For dry air this ratio is $\dfrac{8.2851\dfrac{\mathrm{g}}{\mathrm{mol}}}{R}$

So that the computed value of dT/dz is increased by a factor of $\frac{8.232}{8.2851} = 0.9935$

and the calculated adiabatic lapse rate is 99.35% of the adiabatic lapse rate which ignores non-condensing water.

The point of this problem is first to make them look at that calculation, and second to show that if the water remains as a gas and does not condense or evaporate, then it plays little role and we can rely on the calculations for a dry atmosphere. When water condenses or evaporates, the heat effects are large and produce significant effects in the atmosphere, e.g. thunderclouds and "microbursts".

5.10 (a) The first equation in the problem is simply the perfect gas volume divided by the mass of one mol of water and n mols of air. The second is the density after the water has condensed, and then simplified by taking the volume of the condensed liquid as 0.

Then, for the same system, assuming an adiabatic process we have

$$\Delta h = 0 = nC_P(T_2 - T_1) - \Delta h_{\text{condensation}}$$

$$T_2 = T_1 + \frac{\Delta h_{\text{condensation}}}{nC_P}$$

Then constructing the ratio of ρ_1 to ρ_2 and inserting this value of T_2 we find

$$\frac{\rho_2}{\rho_1} = \frac{(n+1)\frac{RT_1}{P}}{n\frac{R}{P}\left\{T_1 + \frac{\Delta h}{nC_P}\right\}} = \frac{n+1}{n + \frac{\Delta h}{T_1 C_P}}$$

which is less than 1 if $\frac{\Delta h}{T_1 C_P} > 1$. For the values shown in the problem

$$\frac{\Delta h}{T_1 C_P} = \frac{45\frac{\text{kJ}}{\text{mol}} \cdot \frac{1000\text{J}}{\text{kJ}}}{\left(33.8\frac{\text{J}}{\text{mol } K}\right)(273\,\text{K})} = 4.87$$

(b)For water going to ice we can repeat the entire derivation, taking into account that both the water and the ice have negligible volumes (compared to the air) so that

$$\frac{\rho_2}{\rho_1} = \frac{T_1}{T_2} = \frac{1}{1 + \frac{\Delta h_{\text{freezing}}}{nC_P}}$$

which is less than one if $\frac{\Delta h_{\text{freezing}}}{nC_P}$ is greater than zero, which it certainly is.

5.11 The appropriate C_V is shown in the problem, but as shown later the term involving the heat capacity of vapor water is negligible. Then, as shown in the problem, we differentiate Equation (A), finding

$$dV = (1+X)\frac{R}{P}dT - (1+X)\frac{RT}{P^2}dP + \frac{RT}{P}dX$$

we substitute this in the combined equations (B) and (C) to find

$$(1+X)RT\left(\frac{dT}{T} - \frac{dP}{P} + \frac{dX}{(1+X)}\right) = \left(C_{V_{air}} + X\,C_{V_{water}}\right)dT + \Delta u_{cond}dX$$

which we rearrange to

$$dT\left(C_{V_{air}} + XC_{V\,water} + R(1+X)\right) = dP\left((1+X)\frac{RT}{P}\right) - dY\left(\Delta u_{cond} + RT\right)$$

we replace dP from Equation 6.5, and rearrange to Equation E. As a check, if we set $X = 0$ in this equation and observe that for a perfect gas $C_P = C_V + R$, then we find that this is the same as equation 5.9.

Then we set $X = 0$ everywhere except in its derivatives, to find

$$\frac{dT}{dz} = \frac{-gM_{avg}}{C_{P\,air}} - \frac{dX}{dz}\frac{\left(\Delta u_{cond} + RT\right)}{C_{P\,air}}$$

The first term is the dry adiabatic lapse rate (Equation 5.9). In the second term we observe that for a perfect gas $\Delta h = \Delta u + RT$

so that we get Equation 5.11.

5.12 In the afternoon the air contains much more dust than in the morning, because afternoons are windy and nights are calm. Dust is put into the afternoon atmosphere by those winds, and settles out during the calm night. For that reason when the sun is low on the horizon in the afternoon, we look at it through an air mass which has a high concentration of particles. In the morning the air has many fewer particles, so that at sunrise we look at the sun through an air mass with far fewer particles.

These particles scatter and absorb sunlight, so that the intensity of sunlight reaching our eyes is less at sunset than at dawn. (For fear of lawsuits, never tell the students to look directly at the sun. However in most sunset situations one can look directly at the sun with little risk. Please don't sue me!) These particles are more efficient scatterers for blue light than for red (for short waves than for long) so that the light which reaches our eyes at sunset is shifted toward the red end of the spectrum, making sunsets orange or red.

5.13 The prevailing winds in the Hawaiian Islands, and almost anywhere at that latitude are from the northeast, See Figure 5.2. That means that the serious storms all come in from that direction. Those storms often have high winds, driving big waves on shore, and washing away any land at the base of the cliffs. This effect is pretty striking on Kauai, Maui and Hawaii. It is not as striking on Oahu.

5.14 These winds plus the prevailing westerly winds in the temperate zone (plus the current of the Gulf Stream) made possible trade between Europe and North America, using ships which could only sail with the wind behind them. Modern sailing ships (used mostly for racing) can sail very rapidly across the wind, and can make a modest speed upwind. The merchant and military ships of 1500 could sail downwind, and to some extent crosswind, but very poorly upwind.

A ship from Europe would sail south along the coast of Europe to reach the tropics, where the trade winds pushed it across the ocean to Central America and/or the Caribbean Islands. Then it would sail north, visiting the east coast of North America, and ride the prevailing westerlies home to Europe.

5.15 Without the coriolis force. With the coriolis force.

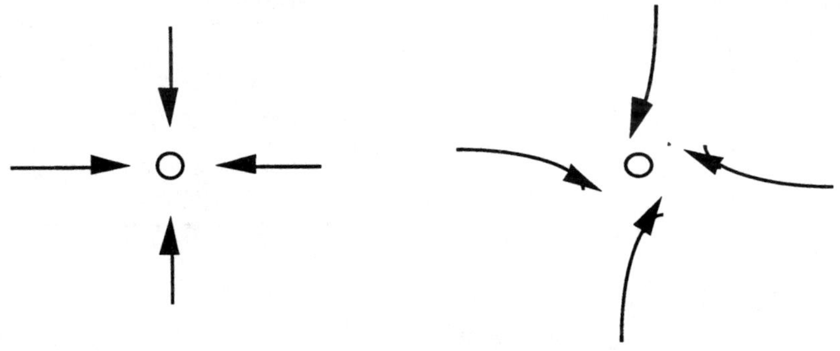

5.16 Applying the frictionless form of Bernoulli's equation from some distance far away to the eye of the hurricane, we find

$$V = \sqrt{\frac{2\Delta P}{\rho}} = \sqrt{\frac{2((1013-850)\,\text{mb})}{1.20\,\frac{\text{kg}}{\text{m}^3}} \cdot \frac{100\text{Pa}}{\text{mb}}} = 165\frac{\text{m}}{\text{s}} = 369\frac{\text{mi}}{\text{hr}}$$

I got this quote out of a book somewhere. I wish I remembered where. The calculation shows that the estimated velocity is low. However this is a frictionless calculation, and if one takes friction into account, the result will be substantially lower. Nonetheless, that low a pressure, at sea level, would correspond to an amazing velocity.

5.17 On a plot of potential temperature vs height, the adiabatic lapse rate is a vertical line. Thus changing from a z-T plot to a z-θ plot simply rotates both lines clockwise until the dashed line, which represents the adiabatic temperature gradient is vertical.

Chapter 6

For all the solutions requiring values of σ_y and σ_z, here and in the text, the values have been calculated using the formulae in Problem 6.16. This has the advantage over chart reading that it gives a unique, reproducible value. One should not assume that it is a more accurate value, since plots and equations for dispersion parameters are all only rough estimates of the true values. If students base their solutions on reading Figures 6.7, 6.8 or 6.9, their results should be close to those shown here, limited by chart reading accuracy.

6.1 Choose the material balance boundaries from the upwind side of the city, $x = 0$ to some arbitrary value of x. Then

$$\text{Flow in} = b\,uWH + qWx$$

$$\text{Flow out} = u\,WHc(x)$$

$$c(x) = b + \frac{qx}{uH}$$

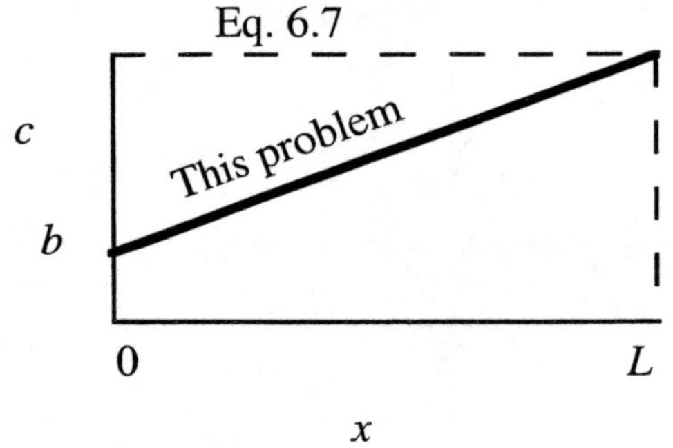

This still has the poor assumption of complete mixing in the vertical. The Hanna-Gifford model, [2]is similar to this model, with that bad assumption replaced by a more reasonable one.

6.2** Use equation 6.7, with the concentration at the downwind side of each strip being the background concentration for the strip downwind of it. Then

$$c_1 = 1\frac{\text{mg}}{\text{m}^3} + \frac{100\text{ g}}{\text{km}^2\cdot\text{s}}\cdot\frac{5\text{ km}}{3\frac{\text{m}}{\text{s}}\cdot 0.4\text{ km}}\cdot\frac{\text{km}^2}{1\text{ E6 m}^2} = 1 + 0.416\frac{\text{mg}}{\text{m}^3}$$

$$c_2 = 1.416 + 500\cdot\frac{2}{3}\cdot\frac{1}{0.5}\cdot\frac{1}{1\text{ E6}} = 1.416 + 0.666 = 2.082$$

$$c_3 = 2.082 + 0.416 = 2.498\frac{\text{mg}}{\text{m}^3}$$

This is also one of the modifications of the simple fixed-box model incorporated into the Hanna-Gifford model [2]. See also Prob. 6.34

6.3 In parts (a) and (b) the assumption that $(z+H)$ is very large makes the exponential using $-(z+H)$ squared practically zero, so we drop it. Then

(a) On the centerline $c = \dfrac{Q}{2\pi u \sigma_y \sigma_z}$

(b) Same as (a), but multiplied by $\exp\left(-0.5\left(\dfrac{y}{\sigma_y}\right)^2\right)$

(c) Here the two exponential terms are identical so that $c = \dfrac{2}{2}\dfrac{Q}{\pi u \sigma_y \sigma_z}\exp\left(-0.5\left(\dfrac{H}{\sigma_z}\right)^2\right)$

(d) $$c = \frac{2}{2}\frac{Q}{\pi u \sigma_y \sigma_z}\cdot\exp\left(-0.5\left(\frac{y}{\sigma_y}\right)^2\right)\cdot\exp\left(-0.5\left(\frac{z}{\sigma_z}\right)^2\right)$$

Here we see some of the logic of the "reflected plume". Because all of the pollutant is at or above the ground surface the concentrations are everywhere exactly twice what they would be if this were a plume high above the ground.

6.4 The corresponding equation for K_x is $\dfrac{1}{2}\sigma_x^2\left(\dfrac{u}{x}\right)$ Then making all the substitutions

$$c = \frac{Q\Delta t}{8\left(\pi\frac{x}{u}\right)^{\frac{3}{2}}\left[\left(\frac{1}{2}\right)^3\left(\frac{u}{x}\right)^3\sigma_x^2\sigma_y^2\sigma_z^2\right]^{\frac{1}{2}}}\cdot\exp-\left[\left(\frac{u}{4x}\right)\left\{\frac{x^2}{\frac{1}{2}\sigma_x^2\frac{u}{x}}+\frac{y^2}{\frac{1}{2}\sigma_y^2\frac{u}{x}}+\frac{z^2}{\frac{1}{2}\sigma_z^2\frac{u}{x}}\right\}\right]$$

$$c = \frac{Q\Delta t}{(2\pi)^{\frac{3}{2}}\sigma_x\sigma_y\sigma_z}\cdot\exp-\left[0.5\left(\frac{x}{\sigma_x}\right)^2+0.5\left(\frac{y}{\sigma_y}\right)^2+0.5\left(\frac{z}{\sigma_z}\right)^2\right]$$

6.5 (a) See the sketch at the right. There is effectively no y direction spreading for the same reason there is no x direction spreading the gaussian plume equation; the dispersion across the boundaries has a net value of zero.

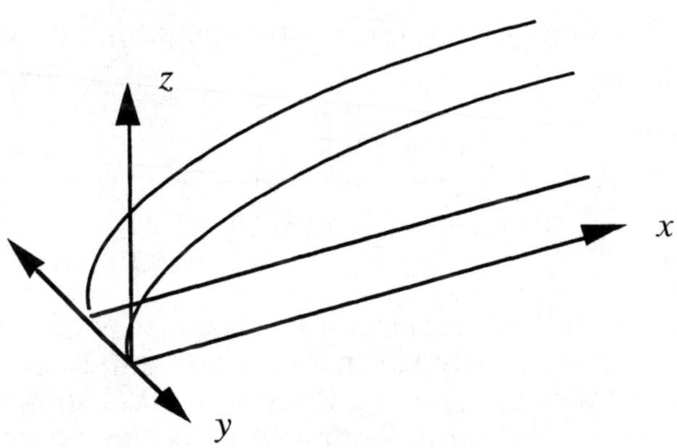

Suppose we put up solid walls parallel to the x axis, as shown in the sketch, allowing no net transport of pollutants across them. Would that change the concentration at any point?

(b) Start with Equation 6.19 for one-dimensional spreading, substitute Equations 6.24 and 6.26.

$$c = \frac{\left(\frac{Q}{L}\right)\frac{1}{u}}{(2\pi)^{\frac{1}{2}}\left(\frac{x}{u}\right)^{\frac{1}{2}}\left(\frac{1}{2}\sigma_z^{\,2}\frac{u}{x}\right)} \cdot \exp\left(-\left(\frac{u}{4x}\right)\cdot\left(\frac{z^2}{\frac{1}{2}\sigma_z^{\,2}\frac{u}{x}}\right)\right) = \frac{\frac{Q}{L}}{(2\pi)^{\frac{1}{2}}u\sigma_z} \cdot \exp\left(-0.5\left(\frac{z}{\sigma_z}\right)^2\right)$$

Then observe that because this is a ground-level source, the concentrations must be multiplied by 2 because of "ground reflection" (See the solution to problem 6.3(d).)

(c) As shown at the right, the strip with width Δx in part (a) now corresponds to the narrower strip with width $x \sin \theta$. The emissions into this strip are the same as into the strip with width Δx in part (a), but the volume of air is less. This is taken care of by placing a $\sin \theta$ in the denominator of the solution in part (a). The resulting formula is the basic formula for most models for estimating the concentration downwind from highways. For $\theta = 0$, this predicts an infinite concentration. Most highway modeling programs have some special modifications to take that into account.

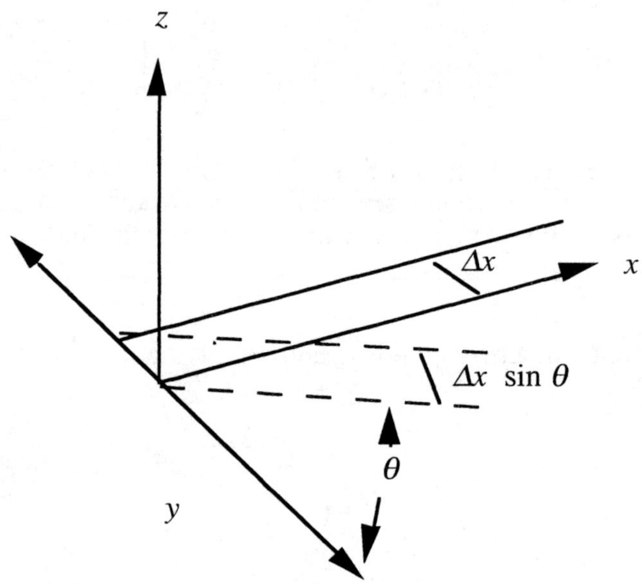

The result in part (a) is also used to estimate concentrations in a narrow canyon, where the spreading is also one-dimensional, and sometimes in the "street canyon" of a city with tall buildings on both sides of a street

---

6.6** Using Equation 6.29; and computing or looking up $\sigma_y = 438$ m and $\sigma_z = 264$ m we have

$$c = \frac{1000\frac{\text{g}}{\text{s}}}{(\pi)\left(3\frac{\text{m}}{\text{s}}\right)(438\text{ m})(264\text{ m})} \cdot \exp\left[-\frac{1}{2}\left(\frac{225\text{ m}}{264\text{ m}}\right)^2\right] = 638\frac{\mu\text{g}}{\text{m}^3}$$

Discussion: I ask the class if this is the peak concentration at ground level, then ask them how one would estimate that peak concentration. That leads to the discussion of Figure 6.9, and then to assigning Problem 6.10. I show in class that the result from Fig. 6.9 is the same as this result (within chart reading accuracy). Using equation 6.31 and the above values one sees that the value one should read from Fig 6.9 is

$$\frac{cu}{Q} = \frac{1}{\pi(438\text{ m})(264\text{ m})} \cdot \exp\left[-\frac{1}{2}\left(\frac{225\text{ m}}{264\text{ m}}\right)^2\right] = 1.912 \cdot \text{E-6}\frac{1}{\text{m}^2}$$

which leads to the same computed value of c. As best I can read Fig. 6.9 the value is 1.90 E-6.

6.7** By simple rearrangement of Equation 6.31

$$H=\left[-2\sigma_z^{\,2}\ln\left(\frac{c\pi u\sigma_y\sigma_z}{Q}\right)\right]^{\frac{1}{2}}=\left[-2\cdot 264\text{m}\cdot\ln\left(\frac{\frac{638\text{ E-6}}{2}\frac{\text{g}}{\text{m}^3}}{1000\frac{\text{g}}{\text{s}}}\cdot\pi\cdot 3\frac{\text{m}}{\text{s}}\cdot 438\text{ m}\cdot 264\text{ m}\right)\right]^{\frac{1}{2}}$$

$= 383.7\text{m}.$ So that $h = H - \Delta h = 383.7 - 75 = 308.7$ m

One can solve the problem in a more general way, in terms of initial and final values as

$$\frac{c_{\text{final}}}{c_{\text{initial}}}=\frac{\exp\left\{-0.5\left(\frac{H_{\text{final}}}{\sigma_z}\right)^2\right\}}{\exp\left\{-0.5\left(\frac{H_{\text{initial}}}{\sigma_z}\right)^2\right\}}=\frac{1}{2}$$

which leads to the same result.

This approach to reducing ground level pollutant concentrations is **strongly discouraged** in current US air pollution law; see the discussion in Chapter 7.

6.8** Here D stability is appropriate, see the note below Table 6.1. Here we asking the distance at which

$$\frac{cu}{Q}=\frac{\left(5\text{ E}-7\frac{\text{g}}{\text{m}^3}\right)\cdot\left(2\frac{\text{m}}{\text{s}}\right)}{\frac{10\text{ g}}{3600\text{ s}}}=3.6\text{ E-4}\frac{1}{\text{m}^2}$$

This is a trial-and-error solution. For an assumed downwind distance of 1 km we read the values of σ_y and σ_z as 32 and 70 m, for which we calculate

$$\frac{cu}{Q}=\frac{1}{\pi\sigma_y\sigma_z}\cdot\exp\left(-0.5\left(\frac{H}{\sigma_z}\right)^2\right)=\frac{1}{\pi\cdot(70\text{ m})\cdot(32\text{ m})}\cdot\exp\left(-0.5\left(\frac{10\text{ m}}{32\text{ m}}\right)^2\right)=1.35\text{ E-4}\frac{1}{\text{m}^2}$$

Subsequent trial and error steps lead show that at a downwind distance of about 0.5 km, the values of σ_y and σ_z are 18.5 and 36 m which gives $cu/Q = 4.12$ E-4/m^2 which is close to the desired answer. Turner's workbook has the equivalent of Figure 6.9 for all six stability categories. Using it for D stability and $H = 10$ m I read the appropriate downwind distance as 0.55 km.

I use this problem to ask the students what compounds we find stinky (i.e. can be detected at the lowest concentrations). The answer is organic compounds containing nitrogen and sulfur. I then ask them why? Our evolutionary ancestors had good reason to want to smell the animals that hunted them, and to smell the animals they hunted. The most individually characterizing compounds which animals of all kinds emit are the organic nitrogen and sulfur compounds. Individual humans differ enough in their individual sulfur and nitrogen compounds that bloodhounds (which have much more acute senses of smell than humans) can track individual humans by the minuscule amounts of those compounds that humans leave behind wherever they go.

When I first gave this problem, the students all knew that Dogpatch was the locale of the Little Abner comic strip. Today's students do not.

6.9 Directly downwind of a ground level source equation 6.31 becomes

$$\frac{cu}{Q} = \frac{1}{\pi \sigma_y \sigma_z};$$

$$Q = c\pi u \sigma_y \sigma_z = 10\ \text{E} - 5\frac{\text{g}}{\text{m}^3} \cdot \pi \cdot 3\frac{\text{m}}{\text{s}} \cdot 213\ \text{m} \cdot 450\ \text{m} = 9.03\ \text{g/s}$$

6.10 One can solve this directly by interpolating a $H = 225$ m line on Figure 6.9, and reading that the maximum occurs at about 2.9 km, with $cu/Q = 2.9 \cdot 10^{-6}/\text{m}^2$. Using the values from Problem 6.16 and my spreadsheet I compute a distance of 2.87 km and $cu/Q = 2.76 \cdot 10^{-6}/\text{m}^2$. Using those values

$$c = \frac{cu}{Q} \cdot \frac{Q}{U} = \frac{2 \cdot 76\ \text{E} - 6}{\text{m}^2} \cdot 1000\frac{\text{g}}{\text{s}} \cdot \frac{\text{s}}{3\ \text{m}} = 921\frac{\mu\text{g}}{\text{m}^3}$$

This is a good discussion problem. Why is this the maximum? Physically? Computationally?

6.11** (a)

$$c = \frac{1000\frac{\text{g}}{\text{s}}}{2\pi \cdot \left(3\frac{\text{m}}{\text{s}}\right) \cdot (438\ \text{m}) \cdot (264\ \text{m})} \cdot \left\{ \exp\left(-0.5\left(\frac{0\ \text{m}}{264\ \text{m}}\right)^2\right) + \exp\left(-0.5\left(\frac{225 + 225}{264}\right)^2\right) \right\} = 566\frac{\mu\text{g}}{\text{m}^3}$$

This shows a lower concentration at the plume centerline than at the ground, which is surprising, but is a correct consequence of the standard gaussian plume (plus ground reflection) assumptions. Because of ground reflection, there is some range of distances downwind at which the highest concentration due to an elevated plume is at the ground, rather than at the plume centerline. See Figure 3-4 of Turner's Work Book.

(b)

$$c = (c \text{ from prob } 6.6) \cdot \exp\left(-\frac{1}{2}\left(\frac{y}{\sigma_z}\right)^2\right)$$

$$= 638\frac{\mu\text{g}}{\text{m}^3} \cdot \exp\left(-\frac{1}{2}\left(\frac{500\ \text{m}}{438\ \text{m}}\right)^2\right) = 356\frac{\mu\text{g}}{\text{m}^3}$$

6.12** From Fig 6.9 I read a downwind distance of 1.9 km and a cu/Q of 6.2 E-6/m^2. Using the values in Problem 6.16 and a spreadsheet I find the maximum at 1.87 km and a cu/Q of 6.17 E-6 /m^2. Thus the answer to part (b) is 1.85 km, and to part (a) is

$$c = \frac{cu}{Q} \cdot \frac{Q}{u} = \frac{\left(6.17\,\mathrm{E\text{-}6}\frac{1}{\mathrm{m}^2}\right) \cdot \left(750\frac{\mathrm{g}}{\mathrm{s}}\right)}{7\frac{\mathrm{m}}{\mathrm{s}}} = 660.8\frac{\mu\mathrm{g}}{\mathrm{m}^3}$$

One should not assume that we know the values on these figures to three significant figures. The numbers in Problem 6.16 allow us to compute to many figures, but the data on which those numbers and these figures are based are certainly not reliable to more than ± 20 or 30%.

This is the point to discuss the shape of Figure 6.9. That figure shows that for the gaussian plume assumptions, raising the stack always lowers the ground level concentrations near the stack, and does not increase the ground level concentration anywhere. The incorrect assumption is that there is no removal of pollutants by the ground. Field tests show that all common atmospheric pollutants are removed at some rate by the ground. Thus raising the stack is certain to lower the nearby concentrations, but by slowing the rate of ground removal it may cause the long distance concentrations to be higher.

6.13 (a) Making that substitution, we have

$$c = \frac{Q}{\pi u}\frac{1}{\alpha\sigma_z^{\,2}} \cdot \exp\left[-0.5\left(\frac{H}{\sigma_z}\right)^2\right]; \quad \frac{dc}{d\sigma_z} = \frac{Q}{\pi u \alpha} \cdot \left(1 - \frac{H^2}{2\sigma_z^{\,2}}\right) \cdot \left(\frac{-2}{\sigma_z^{\,2}}\exp\left(-0.5\left(\frac{H}{\sigma_z}\right)^2\right)\right)$$

setting this = 0 leads to $\sigma_z = \dfrac{H}{\sqrt{2}}$

(b) For H= 20 m, I read the peak on Figure 6.9 at an x of 200 m. From the values in Problem 6.16 that corresponds to a σ_z of 14.3 m. $H/\sqrt{2}$ =14.14 m, so the ratio of these two values is 0.996. For all the curves on Figure 6.9, again based on chart reading for the x at the maximum, this ratio is between 0.995 and 1.03.

Using the comparable charts to Figure 6.9 from Turner's Workbook, I made the same comparison for A and F stabilities. For A the ratio is between 0.95 and 1.34. For F it is between 0.96 and 0.61. For both of these stability categories, the agreement is good at low values of x (where the σ_z plot is practically linear), and much poorer at large values of x, where the plot becomes fan-shaped.

6.14 If one had the equivalent of Figure 6.9 for A stability, which is in Turner's Workbook, one could simply look this up. Without that this problem is a trial and error. Using the values from Problem 6.16 and a spreadsheet it is a fairly quick trial and error. The maximum value of cu/Q is 3.85 E-6/ m^2, which occurs at x = 0.73 km. The corresponding values of the sigma's, which the students could read from Figures 6.7 and 6.8 are 160.8 and 248.6 m. Then we compute that 300 ton/day = 3153 g/s, and

$$c = \frac{cu}{Q} \cdot \frac{Q}{u} = \frac{3.85\,\text{E-}6 \frac{1}{\text{m}^2} \cdot 3153 \frac{\text{g}}{\text{s}}}{3 \frac{\text{m}}{\text{s}}} = 4046 \frac{\mu\text{g}}{\text{m}^3}$$

6.15 One ppm of SO_2 = 1324 μg/m³, so that

$$\frac{cu}{Q} = \frac{1.32\,\text{E-}3 \frac{\text{g}}{\text{m}^3} \cdot 3 \frac{\text{m}}{\text{s}}}{1000 \frac{\text{g}}{\text{s}}} = 3.97\,\text{E-}6 \frac{1}{\text{m}^2}$$

From Fig. 6.9 one reads that one would expect to smell this plume at any distance between 0.6 and 3.7 km downwind. The expected answer, the maximum downwind distance is x 3.7 km.

6.16 For that example

$$\sigma_y = 104 \cdot 0.5^{0.894} = 55.96$$

$$\sigma_y = 61 \cdot 0.5^{0.911} = 32.4$$

This is not a coincidence. The values in the example were computed from these formulae. By direct chart reading from the larger versions of Figures 6.7 and 6.8 in Turner's Workbook, I had previously used the values of 55 and 32.

6.17 Here

$$\frac{Q}{L} = \frac{10{,}000 \text{ vehicle miles / hr}}{\text{mile of highway}} \cdot \frac{3.5 \text{ g}}{\text{vehicle mile}} \cdot \frac{\text{hr}}{3600 \text{ s}}$$

$$= \frac{9.44 \text{ g}}{\text{mile} \cdot \text{s}} \cdot \frac{\text{mile}}{5280 \text{ ft}} \cdot \frac{3.381 \text{ ft}}{\text{m}} = 5.869\,\text{E-}3 \frac{\text{g}}{\text{m} \cdot \text{s}}$$

and for C stability and x = 200 m, we read that σ_z 14 m. Then copying the result from Prob. 6.5, (with ground reflection!)

$$c = \frac{2\frac{Q}{L}}{(2\pi)^{\frac{1}{2}} u \sigma_z} \cdot \exp\left(-0.5 \left(\frac{z}{\sigma_z} \right)^2 \right) = \frac{2 \cdot 5.869\,\text{E-}3 \frac{\text{g}}{\text{m} \cdot \text{s}}}{(2\pi)^{\frac{1}{2}} \cdot 1 \frac{\text{m}}{\text{s}} \cdot 14 \text{ m}} \cdot \exp(-0) = 334 \frac{\mu\text{g}}{\text{m}^3}$$

6.18** Using equation 6.35, for u = 1 m/s we have $80 \frac{\text{ft}}{\text{s}} = 24.4 \frac{\text{m}}{\text{s}}$; $\quad 22 \text{ ft} = 6.71\text{m}$

$$\Delta h = \frac{(24.4)(6.71)}{1} \left(1.5 + 2.68\,\text{E}-3 \cdot 790 \cdot 6.71 \cdot \frac{204}{714} \right) = 910 \text{ m}$$

Then the other values are formed by ratio. Those corresponding to u = 3, 10 and 30 m/s are 303, 91 and 30 m.

In discussing this in class I try to make sure the students recognize that this is a semi-theoretical formula, and that there is no one plume rise formula which is universally applicable. I also try to show them that it is the product of two terms, the first of which is a momentum term (not exactly because it has *VD* where a momentum term would have $\rho V^2 D^2$, but approximately) multiplied by a buoyancy term. I ask the question "If one simply pumped extra ambient air into the flow of combustion gases into a stack, would that increase or decrease the plume rise?" Based on this formula, it would. The dilution air would increase *V* while decreasing *T*. But if we go to large values of dilution air the momentum term continues to increase, while the buoyancy term becomes asymptotic to 1.5. I have heard that that experiment has been tried, and that the results contradict this equation; the observed plume rise decreases as dilution air is added to a stack with a buoyant plume.

One can also discuss large volcanic eruptions, e.g. Mt. St. Helens. That had a "plume rise" of at least 30,000 ft, because it put large amounts of particles in the stratosphere. All large volcanic eruptions do. Most such eruptions are not very hot, so that the buoyancy term is not large, but the velocity and diameter of the "stack" are much, much larger than the values for industrial sources. If one assumes that Mt. St. Helens was a momentum plume, not necessarily a buoyant plume then one might estimate a diameter of 1000 m, an exhaust velocity of 100 m/s and a wind velocity of 3 m/s, making the first term 33 km. This is too high, but it shows that for big volcanic eruptions Holland's formula gives some idea that the plume rise would be into the stratosphere.

6.19 Repeating that problem, for u = 1 m/s

$$\Delta h = \frac{(24.4)\cdot(6.71)}{1}\cdot\left(1.5 + 2.68\ \text{E}-3\cdot 790\cdot\frac{70}{580}\right) = 522\ \text{m}$$

and for the other velocities, 174, 52.2 and 17.4 m. In chapter 11 we will see why wet scrubbers greatly reduce stack gas temperatures, thus reducing plume buoyancy. Thus although modern scrubbers reduce the SO_2 emission rates by about 90%, the nearby reduction in maximum ground level concentration will not be this great, because of the reduced plume rise, and correspondingly increased value of *cu/Q*.

6.20** Using Equation 6.31, with $H = z = y = 0$, we have $c = \dfrac{Q}{u\pi\sigma_y\sigma_z}$

Then, for D stability, we look up the sigmas on Figures 6.7 and 6.8 (or compute them from the equations in Problem 6.16 as I have done here, finding.

x, km	σ_z , m	σ_y, m	c, µg/m^3
0.5	17.8	36.6	206
1	31.5	68.0	64
3	65.4	181.6	12
5	89.1	286.7	5

6.21 Here we retain the term in H in Equation 6.31. Using the spreadsheet we can make up the following table

x, km	$(cu/Q)_{\text{for } H=0}$	$(cu/Q)_{\text{for } H=20\text{m}}$	Ratio of the two
0.5	4.8 E-4	2.6 E-4	0.54
1	1.5 E-4	1.2 E-4	0.8
3	2.68 E-5	2.55 E-5	0.95
5	1.25 E-5	1.215 E-5	0.972

This table shows that for short distance it is quite helpful, but for long distances it doesn't help much.

6.22 Writing Equation 6.30 twice, once for some value of y greater than zero, one once for $y = 0$, and taking the ratio, we find

$$\frac{c(y)}{c(y=0)} = \exp\left[-0.5\left(\frac{y}{\sigma_y}\right)^2\right]$$

For this problem, at x =1 km, σ_y = 68 m, so that

$\pm y$ (m)	$c(y)/c(y{=}0)$
0	1
10	0.99
30	0.914
70	0.606
100	0.361
200	0.016

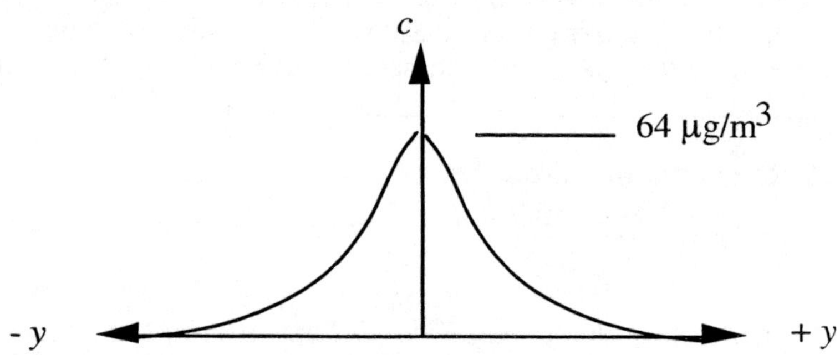

This as close to a Gaussian bell curve as I can do on Macdraw II.

6.23 Using the formulae in Problem 6.16 we calculate that at x = 100 km, σ_z = 6383 m. Substituting in Eq. 6.33, with y = 0, we find

$$\frac{cu}{Q} = \frac{1}{\sqrt{x\pi}\cdot 300\text{ m}\cdot 6383\text{ m}} = 2.08\text{ E-}7\frac{1}{\text{m}^2}$$

As closely as I can read Figure 6.9, this is the value plotted there.

6.24 See Prob. 6.6. Here $\dfrac{cu}{Q} = \dfrac{(30\ \mu\text{g/m}^3)\cdot(3\text{ m/s})}{1000\text{ g/s}} = 9\text{ E}-8\dfrac{1}{\text{m}^2}$

(a) From Fig. 6.9, reading along the L = 1000 m curve, we find this value at x 65 km.

(b) Reading the same figure for L = infinity we follow the curve which is asymptotic to the various H curves, reading x 33 km.

This illustrates the idea of the influence of mixing height on concentrations. At short distances it has no effect, but at long distances it can have a significant effect.

6.25 For u = 1 m/s the plume rise is 200 m, H = 250 m, from Fig 6.9 we read that the maximum value of cu/Q = 2.3 E-6/m^2, and

$$\frac{c}{Q}=\frac{cu}{Q}\cdot\frac{1}{u}=2.3\,\mathrm{E}-6\frac{1}{\mathrm{m}^2}\cdot\frac{1}{1\frac{\mathrm{m}}{\mathrm{s}}}=2.3\,\mathrm{E}-6\frac{\mathrm{s}}{\mathrm{m}^3}$$

Then, we make up a table for various values of u finding that for 2, 3, 4, 5, and 6 m/s the calculated values of $(c/Q)\cdot 10^6$ s/m^3 are , 3, 3.2, 3.5, 3.3, 3.2 (to within chart reading accuracy). So the maximum concentration occurs at a wind speed of about 4 m/s.

This is only approximately right, because as Table 6.1 shows, we would not expect any one stability category to apply over this range of wind speeds,. (D stability applies for all wind speeds under overcast conditions. I would have used that if there were room in the text for the equivalent of Figure 6.9 for D stability.) In spite of this approximation, it is true that the maximum ground level concentrations do not occur at very high or very low values of u, but rather at some intermediate value, because u appears in the concentration formula, but also in the plume rise formula. For a source with no plume rise, the maximum calculated concentration occurs at the lowest wind speed.

6.26 This problem is easy on a spreadsheet, but not very satisfactory as a hand problem. Parts (a) and (b) are easy hand problems, but they only lead up to the interesting part (c) which is not easy by hand.

(a) From Fig. 6.9 we read for H = 250 m we read that the maximum of cu/Q is 2.2 E-6/m^2 and that it occurs at x 2.2 km. Thus the answer to part (a) is

$$c=\frac{cu}{Q}\cdot\frac{Q}{u}=2.2\,\mathrm{E}\text{-}6\frac{1}{\mathrm{m}^2}\cdot\frac{100\ \mathrm{g/s}}{2\ \mathrm{m/s}}=110\frac{\mu\mathrm{g}}{\mathrm{m}^3}$$

and the answer to part (b) is 2.2 km.

(c) The point of this problem is that as u increases, H decreases because the plume rise is proportional to $(1/u)$. The calculated value of c_{max} goes through a maximum with increasing wind speed.

To do it by hand one reads the values of the peak (cu/Q) corresponding to various values of H from Fig. 6.9, then calculates the plume rise corresponding to these, then calculates the corresponding wind speed from

$$u=u_o\frac{\Delta H_o}{\Delta H}=2\frac{\mathrm{m}}{\mathrm{s}}\cdot\frac{150\ \mathrm{m}}{\Delta H}$$ and then computes (c/Q) From $\frac{c}{Q}=\frac{cu}{Q}\cdot\frac{1}{u}$

The result are shown below

H (m)	cu/Q ·1E6 (1/m^2)	Δh, m	u (m/s)	$(c/Q)\cdot$1E6 s/m^3

150	6.2	50	6	1.033
200	3.3	100	3	1.100
250	2.2	150	2	1.100
300	1.5	200	1.5	1.000

If one plots (c/Q) vs x one finds a broad maximum between 2 and 3 km. The limitation is the accuracy of chart reading.

To do it on a spreadsheet one labels the columns, u, plume rise, H, sigma z at c max, x max, sigma y and c. The values of u are chosen from 2 to 6 m/s. The plume rise is generated from the above formula. H = plume rise + 100 m. The sigma z corresponding to the maximum c is computed as shown in Prob. 6.13. Then the distance to this maximum is computed by the formula in Prob. 6.16. Then the sigma y for that distance is computed from the formulae in Prob. 6.16, and finally the c is computed from Eq. 6.31. The resulting table shows that the curve has a shallow maximum at u 3 m/s.

6.27** Using Equation 6.9

$$q = (c - b) \cdot \frac{uH}{L} = \left((35{,}000\ \mathrm{E}-6 - 5000\ \mathrm{E}-6) \frac{\mathrm{g}}{\mathrm{m}^3} \right) \cdot \frac{\left(0.5 \frac{\mathrm{m}}{\mathrm{s}}\right) \cdot (100\ \mathrm{m})}{3000\ \mathrm{m}} = 5\ \mathrm{E}\text{-}4 \frac{\mathrm{g}}{\mathrm{m}^2 \cdot s}$$

6.28 This begins as a gaussian puff problem, see problem 6.4, but the plume quickly spreads up to the mixing layer, and then as the puff travels downwind, it becomes a two-dimensional spreading problem, with spreading in the x and y directions, but not in the vertical. Starting with Eq. 6.20, and taking the viewpoint of someone riding at the center of the cloud, for which $x = y = 0$, and for which $X = (Q\ t/L)$, we write

$$c = \frac{\left(\frac{Q\Delta t}{L}\right)}{4(\pi t)\left(K_x K_y\right)^{\frac{1}{2}}}$$

Then, as assumed in the problem we take $K_x = K_y$, and substitute equations 6.24 and 6.26 to find $c = \dfrac{Q\Delta t}{2\pi\sigma_x\sigma_y L}$

estimating σ_y from the equations in Problem 6.16 for a downwind distance of 1000 km we find $\sigma_y = \sigma_x \approx 104 \cdot (1000)^{0.894} = 50{,}000\ \mathrm{m}$ and

$$c = \frac{10^7\,\mathrm{curies}}{2\pi \cdot (50{,}000\mathrm{m})^2 \cdot (2000m)} = 3.2\ \mathrm{E}\text{-}7 \frac{\mathrm{Ci}}{\mathrm{m}^3} = 3.2\ \mathrm{E}\text{-}10 \frac{\mathrm{Ci}}{\mathrm{L}}$$

$$= 320 \frac{\mathrm{pCi}}{\mathrm{L}}$$

Natural background radiation is about 0.2 pCi/L, so this would be easily detected.

This is certainly a vast simplification of the real situation, but it shows that the expected maximum concentration, 1000 km downwind of the source was about 1600 times the typical radioactivity of the air (See Chapter 15).

W. Klug et. al. *Evaluation of Long Range Atmospheric Transport Models using Environmental Radioactivity data from the Chernobyl Accident*, Elsevier, London and NY, 1992 reports on constructing a concentration-location-time data base, and then letting everyone try their model on it. They concluded that the lagrangian and eulerian models did about equally well, but none were really good. They did not give nearly as much data as I would have liked. They do show that several days after the accident the concentration of iodine in Stockholm was about 2 Bq/m^3 = 54 nCi/m^3= 54 pCi/L which is about an sixth of what I show in this problem!

6.29 We use the last equation in Prob. 6.4. Here this is written in the Lagrangian viewpoint, so the coordinates of the center of the cloud are $x = y = z = 0.00$, so that the exponential terms are all exp 0, and

$$c = \frac{Q\Delta t}{(2\pi)^{\frac{3}{2}}\sigma_x\sigma_y\sigma_z}$$

To account for ground reflection we must multiply the above equation by 2. We can use the equations in Prob. 6.16 to estimate $\sigma_y = 438$ m, $\sigma_z = 264$ m. We have no good way to estimate σ_x , but guessing that it is σ_y is plausible. Thus

$$c = \frac{2 \cdot 100 \text{ g}}{(2\pi)^{\frac{3}{2}} \cdot (438 \text{ m}) \cdot (438 \text{ m}) \cdot (264 \text{ m})} = 2.5 \text{ E-}6 \frac{\text{g}}{\text{m}^3} = 2.5 \frac{\mu\text{g}}{\text{m}^3}$$

The US Army Chemical Corps lists the TWA 's (Time Weighted Averages, concentrations to which workers may be routinely exposed without protective equipment for 40 hours a week) as 0.1, 0.01 and 3 μg/m^3 for the three types of nerve gas (GB, VX and HD). The IDLH (Immediately Dangerous to Life and Health, which corresponds to a 30 minute exposure without symptoms or irreversible health effects) values are 200, 20 and 3 μg/m^3. Thus, the concentration calculated above would be close to the concentration which would trigger the evacuation of a facility, but not be expected to cause health damage.

Sax, *Dangerous Properties of Industrial Materials, 4e* p 1138, Van Nostrand Reinhold, NY, 1975 says that the nerve gases are most dangerous as liquid droplets, because the liquid is easily absorbed through the skin. The vapor is less dangerous because it is less rapidly absorbed. The lethal dose for humans is estimated at 0.01 mg/kg or 1E-8 times the body mass. For a 75 kg person that is 0.75 mg, or about a droplet 1 mm in diameter. If we assume that we breathe in 15 L/minute = 0.015 m^3/minute, then at this concentration we would breathe in about 3.75 E-8 g/minute. This would probably not cause any observable physiologic effect. If one takes a pound of nerve gas and divides it into 0.75 mg doses, one finds one has 6 E5 lethal doses. This is a scary number, but when a Japanese terrorist organization released a pound of two of it in a Tokyo subway station they only killed about 20 people. The 6 E5 number assumes that the pound is divided into 1 mm diameter drops and each drop put on the skin of a separate person. The actual dispersion is much less efficient. Please don't let this text induce you to play with nerve gas; it is dangerous. However, as far as I know we have never had a nerve gas fatality in the USA!

6.30 (a) Originally the dye cloud will be practically spherical, well away from any of the boundaries of the river. As it grows it will reach the top and bottom surfaces of the river, so that its shape will change to more-or-less a vertical cylinder, which increases in radius with time, but which cannot grow in the vertical direction. When the cylinder grows to be as wide as the river it will convert to a more-or-less rectangular band, which spans the river from bank to bank, and which becomes longer (and more dilute) as it flows downstream.

(b) Initially the spreading is three-dimensional. When the cloud reaches the top and bottom surfaces of the river it changes to two-dimensional. When it reaches the banks, it changes to one dimensional. All of this discussion assumes that the downstream velocity of the river is uniform over the river. The slow-moving boundary layers on the bottom and the banks complicate this simple picture.

(c) This is a gaussian puff problem in a parallel-walled canyon below the mixing height. We could also consider the above problem, with the nerve gas release in a tunnel (or a rectangular subway station!)

6.31** Here above the plume the ground reflection term is negligible, so we can use Equation 6.27. rearranged

$$\exp\left(-\frac{1}{2}\left(\frac{z-H}{\sigma_z}\right)^2\right) = \frac{c\,2\pi u\,\sigma_y\sigma_z}{Q}$$

For C stability, 5 km downwind, $\sigma_y = 438$ m, $\sigma_z = 264$ m, so that

$$\exp\left(-\frac{1}{2}\left(\frac{z-H}{\sigma_z}\right)^2\right) = \frac{500\frac{\mu\text{g}}{\text{m}^3}\cdot 2\pi\cdot 3\frac{\text{m}}{\text{s}}\cdot 438\text{ m}\cdot 264\text{ m}}{5000\frac{\text{g}}{\text{s}}} = 0.218$$

$$-\frac{1}{2}\left(\frac{z-H}{\sigma_z}\right)^2 = \ln 0.218 = -1.523$$

$$z-H = 264\text{ m}\sqrt{\frac{-1.523}{-0.5}} = 461\text{ m};\quad z = 461+400 = 816\text{ m}$$

I know of no practical problem of this type. The Kennecott Copper Smelter is about 10 miles from the Salt Lake City Airport, but airport flight paths are parallel to the prevailing winds, so that the plume is parallel to the flight path, not perpendicular to it. On some rare occasion this might be a problem.

6.32 By a material balance on the cabin we have

$$\frac{dc}{dt} = \frac{Q}{V}\left(c_{\text{outside}} - c\right)$$

For this problem Q/V = 10/hr. To find the outside concentration we use equation 6.29 with $z = H$, and for C stability, 5 km downwind, σ_y = 438 m, σ_z = 264 m, which leads to a plume centerline concentration of 480 μg/m³. I took $t = 0$ when $y = 0$ (i.e. when the plane is crossing the centerline of the plume), which means that the time of interest runs from about - 40 to + 40 s. Then with a spreadsheet I computed the value of the various concentrations in two-second increments. So, for example between -38 and -36 s we have an average outside concentration of 0.4 μg/m³ and an increase the cabin concentration of 0.001 μg/m³. Then following the time to +40 s on a spreadsheet we find that the calculated peak concentration in the cabin is about 30 μg/m³, about 26 seconds after the plane passes through the center of the plume.

If one approximates the plume as a square wave, with concentration equal to the centerline concentration out to plus or minus 10 s of flight time (447 m) on each side, then one can solve analytically, finding a concentration in the cabin, when the plane leaves the plume of 25 μg/m³ which is a fair approximation of the spreadsheet solution.

I do not think that this problem has any practical application.

6.33** For x = 0.1 km, F stability, σ_z 2,2 m. Using u = 1 m/s and Equation 6.25, we

find $$K_z = \frac{2\sigma_z^2 u}{x} = \frac{2 \cdot (2{,}2\text{ m})^2 \cdot 1\frac{\text{m}}{\text{s}}}{100\text{ m}} = 0.0968\frac{\text{m}^2}{\text{s}} \cong 968\frac{\text{cm}^2}{\text{s}}$$

This minimum value is of the order of 10,000 times the typical value for the molecular diffusivity of one gas in another.

6.34 For this set of assumptions this is a one dimensional (z direction) spreading problem. For some downwind location one must integrate over the upwind emission areas to find the concentration. The easiest way to begin with the "line source" model in Prob. 6.5. Consider one strip of infinite width in the cross-wind direction and length Δx in the downwind direction. For such a strip we can replace the (Q/L) in the line-source model with $q\Delta x$. You may verify that this is dimensionally correct. Taking ground reflection into account we can say that at some downwind distance α (here we do not use x for downwind distance to avoid confusion with the Δx in the emission term), we can say that the concentration due to one such strip is

$$c = \frac{2q\Delta x}{(2\pi)^{0.5} u\sigma_z}$$

where the σ_z is that corresponding to the downwind distance α. Then to find the contributions of all the upwind strips we sum over all of them, and take the limit as the strips become infinitely small, finding

$$c = \frac{2q}{(2\pi)^{0.5} u}\int_0^{\alpha} \frac{d\alpha}{\sigma_z(\alpha_{receptor} - \alpha_{emitter})}$$

where we show the σ_z varying as the downwind distance. If one then uses a linear approximation for σ_z, e.g. $\sigma_z = b + c\alpha$ with b and c being data-fitting constants, one can substitute and integrate to find

$$c = \frac{2q}{(2\pi)^{0.5} u} \cdot \ln\left(\frac{b + c\alpha}{b}\right)$$

which is very similar to the Hanna-Gifford model, Reference 2 of Chapter 6.

Chapter 7

7.1** (a) Dividing the power computed from Equation 7.4 by that computed from Equation 7.6 and noting that for an ideal gas $Q = \dot{n}R\dfrac{T_1}{P_1}$

we find
$$\frac{Po_{7.4}}{Po_{7.6}} = \frac{P_1\left(\frac{k}{k-1}\right)\cdot\left[\left(1+\frac{\Delta P}{P_1}\right)^{\frac{k-1}{k}}\right]}{\Delta P} = \frac{P_1}{\Delta P}\cdot\left(\frac{k}{k-1}\right)\cdot\left[\left(1+\frac{\Delta P}{P_1}\right)^{\frac{k-1}{k-1}}-1\right]$$

then, taking $k = 1.4$, and
$$\frac{\Delta P}{P_1} = \frac{1}{14.7} = 0.068$$

we find
$$\frac{Po_{7.4}}{Po_{7.6}} = \frac{14.7 \text{ psia}}{1 \text{ psia}}\cdot\left(\frac{1.4}{0.4}\right)\cdot\left[(1+0.068)^{\frac{0.4}{1.4}}-1\right] = 0.976$$
$$\text{Error} = 1 - 0.976 = 0.024 = 2.4\%$$

(b) There are two approximations in Equation 7.6 compared with 7.4. First it assumes that the gas is incompressible, which makes the calculated power higher for Equation 7.6. Second it ignores the increase in temperature of the gas, which works in the opposite sense. The first effect is larger than the second, so using the simpler equation leads to a slight overestimate of the required work or power. This is illustrated by this sketch. The sketch does not show the adiabatic curve, which has a steeper slope (negative) than the adiabatic curve, so that the error area is slightly smaller.

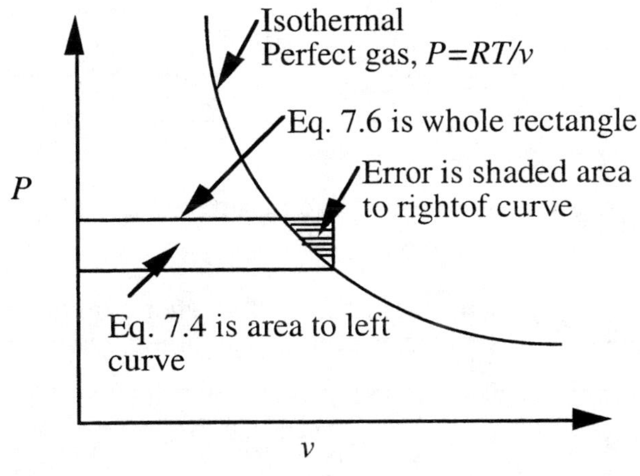

7.2** $p_{\text{overall}} = (p_i)^n = (0.15)^3 = 0.00337;\ \eta_{\text{overall}} = 1 - p_{\text{overall}} = 99.66\%$

This illustrates the utility of penetration as a collector performance measure.

7.3

Size	$\dot{m}_{\text{in}}$	p	$\dot{m}_{\text{out}}$

Large	0.333	$(0.01)^3$	0.0000
Medium	0.333	$(0.25)^3$	0.0052
Small	0.333	$(0.70)^3$	0.1142
		Total	0.1194

$$p = \frac{0.1194}{0.999} = 11.9\%; \quad \eta = 1 - p = 88.1\%$$

7.4** This is an easy trial-and-error solution. First assume that only the small particles contribute to the overall penetration. Then for them

$$p_{\text{small}} = \frac{0.05}{0.33} = 0.150 = (0.70)^n$$

$$n = \frac{\ln 0.150}{\ln 0.70} = 5.3$$

Which leads to an estimate that six collectors in series are needed. Then check by repeating Prob. 7.3 for six collectors, finding

Size	$\dot{m}_{\text{in}}$	p	$\dot{m}_{\text{out}}$
Large	0.333	$(0.01)^6$	0.0000
Medium	0.333	$(0.25)^6$	8.12 E-5
Small	0.333	$(0.70)^6$	0.0391
		Total	0.0392

If one repeats this for five in series one finds $p = 0.056$, slightly above the specified collection efficiency.

This problem shows that when you get to high particle collector efficiencies, only the smallest particles count. The others have all been removed.

7.5

$$p_{\text{small}} = 0.50 \cdot 0.75 = 0.375$$
$$p_{\text{large}} = 0.25 \cdot 0.60 = 0.150$$

$$\sum mp = 0.5 \cdot 0.375 + 0.5 \cdot 0.150 = 0.1875 + 0.075 = 0.2625$$

$$\text{fraction collected} = 1 - 0.2625 = 0.7375 = 73.75\%$$

7.6 The equation to be solved is $p_{\text{overall}} = 0.1 = 0.5 \cdot (1 - \eta_{\text{large}})^5 + 0.5 \cdot (1 - \frac{\eta_{\text{large}}}{3})^5$

I do not think this has an analytical solution. My spreadsheet's "solve" routine shows that the solution is $\eta_{\text{large}} = 0.825661$, from which it follows that $\eta_{\text{small}} = 0.825661 / 3 = 0.275$

7.7 All fuels, CO, coal, coke breeze, natural gas, and waste oils are captured and burned as fuel if their amount and concentration is high enough (Chapter 10).

All solvents are captured and reused if their amount and concentration is high enough, (Chapter 10).

SO_2 is converted to H_2SO_4 and sold when it is available in large amounts at high concentrations. H_2S is captured and converted to elemental sulfur, which is sold in some areas (Chapter 11).

Chlorine gas from chlorine packaging operations is used to make bleach.

The crankcase ventilation control device on autos (Chapter 13) burns the oil and combustion gases which would otherwise be wasted, thus raising the fuel efficiency.

A student who works at the Johnson-Mathey gold processing facility in Salt Lake City told me that their baghouse catch of gold was a significant fraction of their production. I have also heard that a large local jewelry manufacturer captures significant amounts of gold in their waste gas processing devices.

Blast furnace fines in steelmaking are recovered and sintered to be reused.

Catalytic cracker fines are collected and reused.

7.8** Rearranging Equation 6.7 we have

$$Hu = \frac{qL}{c-b} = \frac{\left(2\text{ E}7 \frac{\text{g}}{\text{m}^2\text{hr}} \cdot 5\,\text{mi}\right)}{(10-2)\frac{\text{mg}}{\text{m}^3}} \cdot \frac{1000\,\text{mg}}{\text{g}} \cdot \frac{\text{hr}}{3600\,\text{s}} \cdot \frac{\text{mi}}{1609\,\text{m}} = 2158\frac{\text{m}^2}{\text{s}}$$

We must curtail emissions whenever the *Hu* product is less than this value. From Figure 7.15 it is this low or lower about 12% of the time so that we would need to implement intermittent control 12% of the time.

The point of this problem is to show how intermittent control would work, and to show that it would be impractical for a city, because the required curtailment times are generally too large. However some forms of intermittent control are widely used. Most cities have restrictions on open burning on days with poor dispersion, and many have restrictions on use of wood stoves and fireplaces on days of poor dispersion. The use of oxygenated fuels in winter is a large-scale example of this same thing! (The city fathers in Los Angeles asked people to curtail their driving during the Olympic Games there, to facilitate the games by minimizing traffic jams. By all reports people did drive much less for those few days, so that this kind of curtailment is practical if there is widespread enough public support for it.)

7.9 a = \$41,000, b = \$8.75/ft^2, c = 1.00. If we base it on volumetric flow rate, then a and c will be unchanged, and $b = \frac{\$8.75}{\text{ft}^2} \cdot \frac{\text{min}}{2\text{ ft}} = \frac{\$4.375}{\text{ft}^3/\text{min}}$

7.10 See Example 7.7. This is a simple rework of that, used as a vehicle for discussing combustible limits.

LEL = 0.51 $\cdot$4.02% = 2.05%, UEL = 2.83 $\cdot$4.02% = 11.38 %

There are a variety of reported values of explosive limits. Table 7.1 is widely reprinted and assumed to be authoritative, but, for example *NFPA 58* lists the explosive limits of commercial propane as 2.15 to 9.6 %. There is some difference between commercial propane and the pure propane in Table 7.1, but probably not enough to cause this difference. The difference is most likely due to the difficulty of measuring these limits, and hence the differing experimental results in the literature.

The instructor should talk about gasoline spilled on ground and ignited. At the ground surface there is no flame, because the mixture is too rich. High above the flame there is no combustion, because it is too lean. The flames seek out the places where it is neither too rich nor too lean, and live there. As a general proposition, with exceptions, the headspace mixture in a gasoline tank is too rich to burn, that in a diesel fuel tank too lean, that in a jet fuel tank just right. For that reason jet fuel handling is scarier than gasoline or diesel handling. Ask the students to watch a jet airliner being refueled. The first thing done is to attach the grounding and bonding clamps, the last thing done is to remove them. That is more important with jet fuel than gasoline, for this reason.

One can ask the students why, in Table 7.1, as the molecular weight goes up, the stoichiometric mixture ratio goes down. On a weight instead of mole basis, the stoichiometric ratio is practically the same for any class of fuels, e.g. alkanes (See Prob. 7.11). One can also make sure that the students see that while upper and lower combustible limits, expressed as mole % vary widely, when they are expressed as a fraction of the stoichiometric mixture they do not vary nearly as much, see the next problem.

In the movies and novels, shooting a bullet through a gasoline tank causes an explosion. In real life it does not. I have personally put some liquid gasoline in old, sealed bottles, shaken them to saturate the headspace air, and shot bullets through them. They do not ignite. You can safely try that yourself. So much for the technical accuracy of movies! I still would like to be James Bond; it works for him!

7.11 For methane

$$\begin{pmatrix}\text{stoichiometric}\\ \text{weight fraction}\end{pmatrix} = \frac{0.0964 \cdot 16}{0.0964 \cdot 16 + (1 - 0.0964) \cdot 29} = 0.0556$$

Proceeding the same way (on a spreadsheet!) we can make up the following table.

Compound	*M*, g/mol	Stoic. mol %	Stoich wt. fr	LEL wt fr	UEL wt fr
methane	16	9.46	0.056	0.026	0.082
ethene	28	6.52	0.063	0.027	0.283
ethane	30	5.64	0.058	0.030	0.140
propane	44.1	4.02	0.060	0.031	0.150
1-butene	56.1	3.37	0.063	0.035	0.190
n-butane	58.1	3.12	0.061	0.034	0.173
n-pentane	72.1	2.55	0.061	0.034	0.188
benzene	78	2.71	0.070	0.039	0.196
cyclohexane	84.2	2.27	0.063	0.031	0.211
n-hexane	86	2.16	0.061	0.032	0.206
toluene	92.1	2.27	0.069	0.031	0.190
n-heptane	100	1.87	0.062	0.034	0.227

n-octane	114	1.65	0.062	0.033	0.218
n-nonane	128.3	1.47	0.062	0.030	0.221
n-decane	142	1.33	0.062	0.029	0.189

For simple hydrocarbons the stoichiometric wt. fraction is about 6%. Methane, benzene and toluene are slightly different because their H/C ratios are different from the others in the table, but the difference is not great. The same remarks apply to the lower and upper combustible limits by weight.

7.12 Air-fuel ratios are almost always stated in weight terms, e.g. lb air/ lb fuel. Most other combustion calculations are in molar terms, e.g. mol air/mol fuel. Thus converting from one to the other is mostly fussing with molecular weights. One mole of oxygen weighs 32 grams, together with its associated nitrogen from the air it weighs

$$32\left(1+\frac{0.79}{0.21}\cdot\frac{28}{32}\right)=137.33\frac{\text{g air}}{\text{mol O}_2}$$

(a)

$$\text{C}+\text{O}_2\rightarrow\text{CO}_2;\quad \frac{A}{F}=\frac{137.33}{12}=11.44$$

$$\text{H}_2+\frac{1}{2}\text{O}_2\rightarrow\text{H}_2\text{O};\quad \frac{A}{F}=\frac{\frac{137.33}{2}}{z}=34.33$$

(b)

$$\text{C}_x\text{H}_y+\left(x+\frac{y}{4}\right)\text{O}_2\rightarrow x\text{CO}_2+\frac{y}{2}\text{H}_2\text{O}$$

$$\frac{A}{F}=\frac{\left(x+\frac{y}{4}\right)137.33}{12x+y}=\frac{137.33\left(1+\frac{1}{4}\frac{x}{y}\right)}{12+\frac{y}{x}}$$

One may check this by substituting the *y/x* for methane (= 4) and finding 17.2. The difference between this value and the 17.3 in the example is round-off error in the molecular weights. As the *M* of hydrocarbons increases, *y/x* approaches 2.0.

7.13 LEL = 0.45·29.5% = 13.3 mol %
(a) The intended flow diagram is sketched below

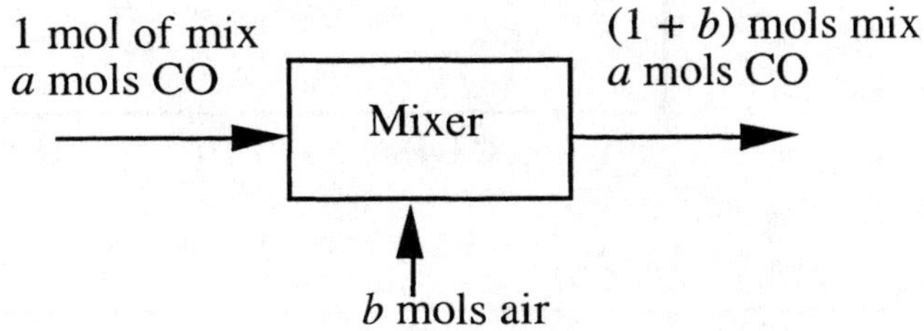

The mol fraction CO in the mix is $y_{CO} = \dfrac{a}{1+b}$

The required amount of oxygen to burn all the CO is $b = \dfrac{\frac{a}{2}}{0.21\dfrac{\text{mol } O_2}{\text{mol air}}} = 0.238a$

$$0.133 = \frac{a}{1 + 0.238a}$$

so that

$$a = \frac{0.133}{1 - (0.133)(0.238)} = 0.194 = 19.4\%$$

(b) The processes are metallurgical processes in which C takes O away from metal oxides to produce the metal (iron, aluminum, zinc, ferroalloys, phosphorus) and partial oxidation processes like catalytic cracker catalyst regeneration.

7.14 At the lean limit there is a surplus of oxygen; replacing the nitrogen with oxygen does not have much effect. At the rich limit there is not enough oxygen to react with all the fuel, and the dilution by the fuel puts out the flame. Adding oxygen at the rich limit moves the rich limit to much higher values. See Fig. 7.4.

The safety consequence is that oxygen atmospheres are very dangerous. NASA thought they could control the hazards and save cost and weight by having our space vehicles operate with an oxygen atmosphere. After they burned three astronauts to death in an all-oxygen cabin fire they decided that even they weren't so clever that they could control all the hazards, and they went over to an oxygen-nitrogen mixture. Often a welder will be working in some enclosed space where the welding fumes make it hard to see or breathe. He or she will ask an assistant to provide some ventilation. The assistant sees a tank of compressed gas, thinks it contains air (or does not know the difference between oxygen and air), and uses it to ventilate the space. In a pure oxygen environment hair burns, skin burns, most clothing burns explosively. The welder normally does not survive.

7.15 See Fig. 7.4 The plot probably looks somewhat like the sketch at the right. The further one departs from stoichiometric, the higher the spontaneous ignition temperature becomes.

The spontaneous ignition temperature is that at which a mixture of the fuel and air ignites without some local ignition source, like a spark.

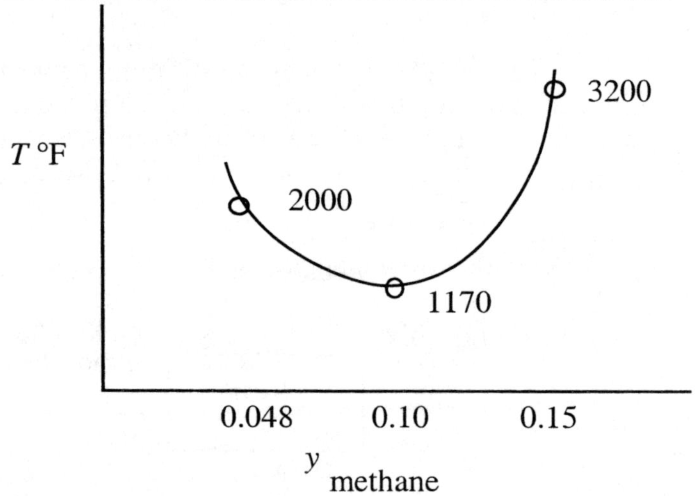

7.16 $\frac{T_2}{T_1} = \left(\frac{V_1}{V_2}\right)^{k-1} = (7)^{0.4} = 2.18$

$T_2 = 530^\circ R \cdot 2.18 = 1155^\circ R = 694^\circ F; \ \Delta T = 694 - 70 = 624^\circ F$

$\frac{T_2}{T_1} = \left(\frac{V_1}{V_2}\right)^{k-1} = (7)^{0.4} = 2.18$

$T_2 = 530^\circ R \cdot 2.18 = 1155^\circ R = 694^\circ F; \ \Delta T = 694 - 70 = 624^\circ F$

Many modern autos have compression ratios of 10, for which the same calculation leads to an adiabatic temperature rise of 1227°F. The actual rise is less than this because of heat transfer to the cylinder walls, head and piston, and because as the temperature increases the heat capacity increases, so that *k* decreases. See the discussion in Chapter 13.

7.17 See Example 7.9. We choose as our basis 1 mol of methane, which (at 20% excess air) produces 12.56 mols of exhaust gas. The heat loss to the stack gas in Figure 7.8 a, as a fraction of the heat released by the methane is

$$\frac{\text{Heat loss to stack}}{\text{Heat input}} = \frac{\left(\dot{n} C_P \Delta T\right)}{\left(\dot{n}_{\text{fuel}} \Delta H_{combustion}\right)}$$

$$= \left(12.56 \frac{\text{mol gases}}{\text{mol methane}}\right) \cdot \left(\frac{7\,\text{Btu}}{\text{lb mol °F}}\right) \cdot \frac{(750-70)\text{°F}}{\left(21{,}502 \cdot 16 \frac{\text{Btu}}{\text{lb mol}}\right)} = 0.1737$$

In Figure 7.8 b the corresponding quantity is

$$\frac{\text{Heat loss to stack}}{\text{Heat input}} = \frac{(12.56)\cdot(7)\cdot(250-70)}{(21{,}502)\cdot(16)} = 0.046$$

So that the increase in thermal efficiency is $\Delta n = 0.1737 - 0.046 = 0.128 = 12.8\%$

One could also say $\Delta n = \frac{0.128}{1 - 0.1737} = 15.5\%$

This is a substantial enough saving that most very large industrial scale furnaces do this.

7.18 $n_{\text{total out}} = n_{\text{total in}} - n_{\text{O}_2\text{consumed}} + \left(n_{\text{CO}_2} + n_{\text{H}_2\text{O}}\right)_{\text{produced}}$

$$= n_{\text{Stoich}} \left(\frac{1+E}{0.21}\right)(1+X) - n_{\text{stoich}} + x + \frac{y}{2}$$

Adding and factoring produces Equation 7.17.

7.19** $\dot{n}_{\text{total out}} = 44.24\dfrac{\text{mols}}{100\,\text{g dry coal}}\cdot\dfrac{1000\,\text{kg coal}}{\text{hr}}\cdot\dfrac{1000\,\text{g}}{\text{kg}}\cdot\dfrac{\text{hr}}{60\ \text{min}} = 7.37\ \text{E3}\ \dfrac{\text{mols}}{\text{min}}$

$$Q = 7.37\ \text{E3}\dfrac{\text{mol}}{\text{min}}\cdot\dfrac{\text{lb mol}}{454\,\text{mol}}\cdot\dfrac{386\ \text{scf}}{\text{lb mol}} = 6{,}266\,\text{scfm}$$

$$= 6266\cdot\left(\dfrac{860^\circ\text{R}}{528^\circ\text{R}}\right) = 10{,}206\,\text{acfm}$$

7.20 Choose the same basis as in Example 7.10, 100 g (dry weight) of coal. Then the amount of contained moisture is

$$H_2O = 100\,\text{g}\cdot\dfrac{0.05}{0.95} = 5.623\ \text{g}\cdot\dfrac{\text{mol}}{18\,\text{g}} = 0.292\,\text{mols}$$

and

$$n_{H_2O\ \text{total}} = 2.99 + 0.292 = 3.28\,\text{mols}$$

$$n_{\text{total out}} = 44.24 + 0.292 = 44.53\,\text{mols}$$

all other values of n_i are unchanged.

$$y_{SO_2} = \dfrac{0.05}{44.53} = 1123\ \text{ppm}$$

7.21 (a) $Q = \dot{m}v = \dfrac{\dot{m}}{\rho} = 1000\dfrac{\text{lbm}}{\text{min}}\cdot\dfrac{\text{ft}^3}{0.075\ \text{lbm}} = 1.333\ \text{E4 scfm}$

(b) $\text{acfm} = \text{scfm}\cdot\dfrac{2660^\circ\text{R}}{528^\circ\text{R}} = 6.715\ \text{E4 acfm}$

The point of this is to get them to look at scfm, acfm. Discuss how one would change that. (T and P). Ask them what the acfm would be if this were in Salt Lake City, where the atmospheric pressure is 12.5 psi. (Answer, [14.7/12.5 = 1.17] times the above answer). Alas, there is no common agreement on the temperature at which we figure scfm. In this book I use 20°C = 68°F (see endpapers). Many use 0°C = 32°F, or 60°F = 15.55 °C. Many EPA regulations are based on 25°C = 77°F.

7.22 $\dot{n}_{\text{water}} = 0.047\dfrac{\text{lb}}{\text{lb}}\cdot\dfrac{100\,\text{lb gas}}{\text{min}}\cdot\dfrac{\text{lbmol}}{18\,\text{lbm}} = 0.261\dfrac{\text{lbmol}}{\text{min}}$

$$\dot{n}_{\text{water}} = 0.047\dfrac{\text{lb}}{\text{lb}}\cdot\dfrac{100\,\text{lb gas}}{\text{min}}\cdot\dfrac{\text{lbmol}}{18\,\text{lbm}} = 0.261\dfrac{\text{lbmol}}{\text{min}}$$

$$Q = (3.448 + 0.261)\dfrac{\text{lbmol}}{\text{min}}\cdot\dfrac{386\,\text{scf}}{\text{lbmol}} = 1{,}432\ \text{scfm}\cdot\dfrac{760^\circ\text{R}}{528^\circ\text{R}} = 2061\ \text{acfm}$$

Again, this is a simple exercise in scfm, acfm.

7.23 Start with example 7.9, with the same basis, 1 mol of methane. Then

$$n_{\text{stoich}} = 2.0; \quad n_{\text{dry}} = 2 \cdot \left(\frac{1+0}{0.21}\right) = 9.524 \text{ mols}$$

$$n_{\text{total}} = 9.524 \cdot (1+0.0116) = 9.634 \text{ mols}; \quad n_{\text{H}_2\text{O}} = 2 + 0.0116 \cdot (9.524) = 2.1105$$

$$n_{\text{total out}} = 1+2+2 \cdot \left[\frac{1.0116}{0.21} - 1\right] = 10.634 \text{ mols}; \quad y_{\text{water}} = \frac{2.1105}{10.634} = 0.198$$

$$p_{\text{water}} = 0.198 \cdot (14.7 \text{psia}) = 0.291 \text{psia}$$

From any steam table we find that a vapor pressure of 2.91 psia corresponds to a saturation temperature of 140°F.

7.24 This problem assumes knowledge of psychrometric charts and the defined quantities on them. Chemical Engineering sophomores learn that. If you assign this problem to a class with students from other disciplines, give them some background on dew point first.**

(a) From that example $X = 0.0473$ (lb water/lb dry air) One can simply look the dew point up on a psychrometric chart. Using the high temperature chart on page 12.5 of *Perry's Chemical Engineers' Handbook, 6e*, one reads a dew point of 103 °F. If one does not have such a chart one can compute the partial pressure of water in the gas as follows

$$X = 0.047 \frac{\text{lb}}{\text{lb}} \cdot \frac{29}{18} = 0.076 \frac{\text{mol water}}{\text{mol air}}$$

$$\text{at the dew point } Y = \frac{p}{P-p}; \; p = \frac{0.076 \cdot 14.7 \text{ psia}}{1+0.076} = 0.104 \text{psia}$$

The dew point temperature is that temperature at which the vapor pressure of the pure liquid equals this partial pressure. From any steam table one may look up that temperature, again finding 103 °F. One could also compute this using the Antoine equations constants in Appendix A.

(b)This requires a trial and error, even with a psychrometric chart. As a first guess try T= 125°F. We can compute the added water as

$$\Delta x(\text{weight}) = \frac{-2.37 \cdot 10^{-4}}{^\circ\text{F}} \cdot (125 - 300)^\circ\text{F} = 0.0415 \frac{\text{lb}}{\text{lb}}$$

and the final water as

$$x(\text{weight}) = 0.0473 + 0.0415 = 0.0888 \frac{\text{lb}}{\text{lb}} = 0.143 \frac{\text{mol}}{\text{mol}}$$

Then the vapor pressure is $p = \frac{0.143}{1.143} \cdot 14.7 \text{psia} = 1.89 \text{psia}$

and the corresponding dew point is 124°F.

The difference is only 1 deg, much smaller than the required 25°F. The assumed outlet temperature is much too low. By repeated trial and error one finds that for a gas temperature of about 147 °F the dew point is about 122°F, so that this is the required solution. At this solution we may add only 0.0367 lb/lb of water, compared to the 0.0415 computed above. This shows that the dew point is very sensitive to the amount of water added!

(c) As sketched at the right, the dew point corresponding to any point on an absolute humidity ratio vs temperature plot is found by drawing a horizontal line from the point to the saturation curve, and reading the corresponding temperature. For the process in part (b) one follows the adiabatic saturation (constant enthalpy) up diagonally to the left, to find the point at which the horizontal distance between that curve and the saturation curve is 25°F.

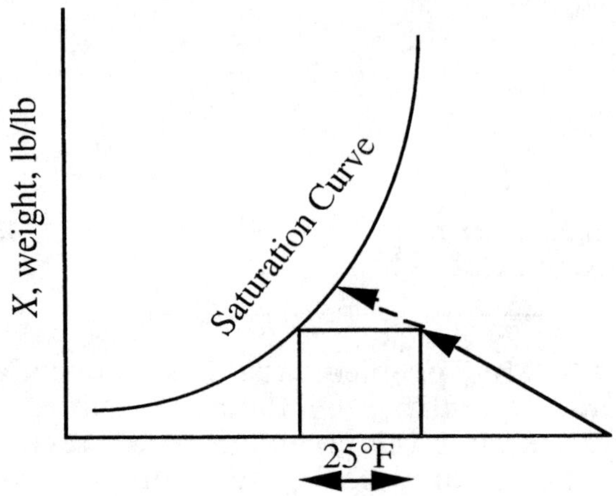

7.25**

$$\frac{\lambda + C_P\Delta T}{\lambda} = 1 + \frac{C_P\Delta T}{\lambda}$$

$$\frac{C_P\Delta T}{\lambda} = \frac{0.25\dfrac{\text{Btu}}{\text{lb}\cdot{}^\circ\text{F}}\left(-200^\circ F\right)}{1055\dfrac{\text{Btu}}{\text{lb}}} = -\,0.047 = -\,4.7\%$$

This approximation is commonly made, because the resulting error is small.

7.26 See Example 7.5. Choose as your basis 1 mol of fuel.

(a) $$\text{mol}\,H_2O = \frac{y}{2} + n_{\text{stoich}}\cdot\left(\frac{X\cdot(1+E)}{0.21}\right)$$

$$\text{Total mols} = x + \frac{y}{2} + n_{\text{stoich}}\cdot\left[\frac{(1+E)}{0.21}(1+X) - 1\right]; \qquad n_{\text{stoich}} = x + \frac{y}{4}$$

$$\text{mol fr}\,H_2O = \frac{\dfrac{y}{2} + n_{\text{stoich}}\cdot\left[\dfrac{X(1+E)}{0.21}\right]}{x + \dfrac{y}{2} + n_{\text{stoich}}\cdot\left\{\dfrac{(1+E)}{0.21}(1+X) - 1\right\}}$$

substitute for n_{stoich} and multiply top and bottom by $2/x$ to get

$$\text{mol fr}\,H_2O = \frac{\dfrac{y}{x} + \left(2 + 0.5\dfrac{y}{x}\right)\cdot\left(\dfrac{X}{0.21}(1+E)\right)}{2 + \dfrac{y}{x} + \left(2 + 0.5\dfrac{y}{x}\right)\cdot\left\{\dfrac{(1+X)}{0.21}(1+E) - 1\right\}}$$

(b) to save writing let $\frac{y}{x} = \alpha$ and let $\frac{X}{0.21}(1+E) = \beta$ and let $\left\{\frac{(1+X)}{0.21}(1+E) - 1\right\} = \gamma$

and let mol fr $H_2O = y_w$

then

$$y_w = \frac{\alpha + (2 + 0.5\alpha)\cdot\beta}{2 + \alpha + (2 + 0.5\alpha)\cdot\gamma}$$

$$\alpha = \frac{2\cdot(y_w\cdot(1+\gamma)) - \beta}{1 + 0.5\beta - y_w\cdot(1 + 0.5\gamma)}$$

for $X = E = 0$, we have $\beta = 0; \quad \gamma = \frac{1}{0.21} - 1 = 3.76$

$$\alpha = \frac{y}{x} = \frac{2\,y_w\cdot(1+3.76)}{1 - y_w(1 + 0.5\cdot 3.76)} = \frac{9.52y_w}{1 - 2.88y_w}$$

and for $y_{water} = 0.11$ this corresponds to

$$\alpha = \frac{y}{x} = \frac{9.52\cdot 0.11}{1 - 2.88\cdot(0.11)} = 1.53$$

(c) This is too high to be a coal, but slightly low for a residual oil.

(d) For this plausible choice of X and E ,we have $\beta = \frac{(0.0116)\cdot(1.2)}{0.21} = 0.0663$

$$\gamma = \left\{\frac{(1.0116)\cdot(1.2)}{0.21} - 1\right\} = 4.781$$

$$\alpha = \frac{y}{x} = \frac{2(0.11\cdot(1+4.781) - 0.0663)}{1 + 0.5\cdot(0.0663) - 0.11\cdot(1 + 0.5\cdot 4.781)} = 1.72$$

which is a plausible value for a residual oil. Thus it is most likely that Figure 7.12 is intended to represent the acid dew point situation for the burning of residual oils.

7.27 Repeating Example 7.10 for this coal, we first note that the values are given per weight of wet coal. The simplest way to proceed is to choose as our basis 100 g of wet coal. Then one finds

$$n_C = \frac{68.3}{12} = 5.96; \; n_H = 4.90; \; n_S = \frac{0.6}{32} = 0.019, \; n_{N_2} = \frac{0.8}{28} = 0.029$$

$$n_{O_2} = \frac{10.2}{32} = 0.319; \; n_{\text{stoich}} = 5.69 + \frac{4.90}{4} + 0.019 - 0.319 = 6.62 \frac{\text{mol}}{100 \text{ g wet coal}}$$

$$n_{\text{Total}} = \frac{(6.62)\cdot(1.20)\cdot(1.0116)}{0.21} = 39.26; \; n_{H_2O} = \frac{6.4}{18} + \frac{4.9}{2} + 39.26\frac{(0.0116)}{1.0116} = 3.24$$

$$n_{\text{Total out}} = 5.69 + \frac{4.90}{2} + 0.019 + 0.29 + 6.62\cdot\left[\frac{(1.2)\cdot(1.0116)}{0.21} - 1\right] + 0.355 = 40.19$$

$$y_{SO_2} = \frac{(0.019)}{40.19} = 472 \text{ ppm}$$

The SO_2 mol fraction is 472 ppm, converting 5% of this produces 24 ppm. From Figure 7.10 this corresponds to an acid dewpoint of $135°C = 275°F$. This is only an estimate, and perhaps a high one, because it is for a gas with less water than the 11% in that figure. Still, it shows the high value observed for the acid dew point.

In my personal experience I have seen three examples in which the equipment designers underestimated the acid dew point problem with disastrous results. My tract house in San Rafael Calif had a radiant heating hot water system. The water heater for it (called a boiler in the plumbing industry) lost its copper heating coil because of acid dew point corrosion. There isn't much sulfur in natural gas, but there is enough in the mercaptan odorant that the condensate on that coil corroded through the coil.

I got to visit the Paradise Ky. TVA powerplant when it was shutdown for an overhaul, mostly because acid dewpoint corrosion had destroyed the air preheater.

I have investigated a residential fatality in which a "high efficiency" forced air furnace had its heat exchanger perforated by acid dewpoint corrosion, allowing combustion chamber CO to enter the house and kill two people.

You can see why I emphasize this topic more than most authors do.

There is some disagreement about the proper estimates of acid dewpoint, mostly because it is difficult to estimate the fractional conversion of SO_2 to H_2SO_4. Figure 4 page 13-4 of *Steam, its Generation and Use*, Babcock and Wilcox, N.Y., 1972 shows the acid dew point design guidelines of this major supplier of boilers. They show much higher dew points for residual oil (about 240°F) than for coals (about 160°F), for a wide range of sulfur contents. I believe this reflects the fact that most residual oils contain some heavy metals in the ash, which act as catalysts for the conversion of SO_2 to H_2SO_4 so that the fractional conversion is higher for residual oils than for coal. Probably on an acid dew point vs mole fraction H_2SO_4 plot their values would agree with Figure 7.12.

7.28 $V = L\cdot A = 1.38 \text{ E-3 m}^3$

(a) $$\rho = 1.20\frac{\text{kg}}{\text{m}^3}\cdot\frac{293 \text{ K}}{(273+590)\text{K}} = 0.0407\frac{\text{kg}}{\text{m}^3}$$

$$Q = \frac{\dot{m}}{\rho} = \frac{136 \text{ kg/h}}{0.407 \text{ kg/m}^3} = 334.2\frac{\text{m}^3}{\text{h}} = 0.0928\frac{\text{m}^3}{\text{h}}$$

$$\text{VHSV} = \frac{Q}{V} = \frac{334.2\dfrac{\text{m}^3}{\text{h}}}{1.387\ \text{E-3 m}^3} = 2.409\text{E5}\frac{1}{\text{h}}$$

(b) For a square channel

$$\text{HR} = \frac{\text{area}}{\text{perimeter}} = \frac{4D}{D^2} = \frac{D}{4} = \frac{1.5\ \text{mm}}{4} = 0.375\ \text{mm}$$

$$\frac{A}{V} = \frac{(4 \cdot D \cdot L)}{D^2 \cdot L / 0.62} = \frac{0.62 \cdot 4}{D} = \frac{0.62 \cdot 4}{1.5\ \text{mm}} = 1653\frac{1}{\text{m}} = 1654\frac{\text{m}^2}{\text{m}^3}$$

(c) $$V = \frac{Q}{A} = \frac{0.0928\ \text{m/s}}{0.62 \cdot 0.0182\ \text{m}^2} = 8.22\frac{\text{m}}{\text{s}}$$

From the empirical correlation for the viscosity of air in the "Common values..." we find that at this temperature the viscosity of air is 3.84 E-5 Pa·s, so that

$$\mathcal{R} = \frac{4\text{HR} \cdot V\rho}{\mu} = \frac{4 \cdot (3.75\ \text{E-4 m}) \cdot \left(8.22\dfrac{\text{m}}{\text{s}}\right) \cdot \left(0.407\dfrac{\text{kg}}{\text{m}^3}\right)}{(3.84\ \text{E-5 Pa} \cdot \text{s})} = 130.68$$

$$\left(K_c + K_e + A_g f\frac{L}{\text{HR}}\right) = \left(0.15 + 0.15 + 1.1 \cdot \frac{14.23}{131} \cdot \frac{0.0762\ \text{m}}{0.000375\ \text{m}}\right) = 24.63$$

$$\Delta P = \frac{\left(0.407\dfrac{\text{kg}}{\text{m}^3}\right) \cdot \left(8.22\dfrac{\text{m}}{\text{s}}\right)^2 \cdot (24.63)}{2} = 339\ \text{Pa} = 0.049\ \text{psi}$$

(d)

$$\begin{pmatrix}\text{Number}\\ \text{of beads}\end{pmatrix} = N = \frac{(1-\varepsilon) \cdot \text{Bed volume}}{\text{Bead volume}} = \frac{(1-0.38) \cdot 1.387\ \text{E-3 m}^3}{\dfrac{\pi}{4} \cdot (3.2\ \text{E-3 m})^3} = 3.34\ \text{E4}$$

$$\text{Bead Surface} = 4\pi N D^2 = 4\pi \cdot (3.34\ \text{E4}) \cdot (3.3\ \text{E-3 m})^2 = 4.30\ \text{m}^2$$

$$\frac{A}{V} = \frac{4.30\ \text{m}^2}{1.387\ \text{E-3 m}^3} = 3100\frac{\text{m}^2}{\text{m}^3}$$

This is slightly less than twice the value for the monolith, certainly the same order of magnitude. For a bed of spheres, any fluid mechanics books shows that

$$\text{HR} = \frac{D}{6} \cdot \frac{\varepsilon}{1-\varepsilon} = \frac{3.2\ \text{mm}}{6} \cdot \frac{0.38}{1-0.38} = 0.326\ \text{mm}$$

(e) Here the interstitial velocity is

$$V_{\text{interstitial}} = \frac{Q}{A} = \frac{0.0928\dfrac{\text{m}^3}{\text{s}}}{(0.38) \cdot (0.0182\ \text{m}^2)} = 13.41\frac{\text{m}}{\text{s}}$$

$$\mathcal{R} = \frac{4\text{HR}\cdot V\rho}{\mu} = \frac{4\cdot(3.26\text{ E - 4 m})\cdot\left(813.41\frac{\text{m}}{\text{s}}\right)\cdot\left(0.407\frac{\text{kg}}{\text{m}^3}\right)}{(3.84\text{ E - 5 Pa}\cdot\text{s})} = 185.3$$

Based on Figure 6 of [24], f 0.8 where the entrance, exit and flow development values have been included in the correlation, so that

$$\Delta P = f\frac{L}{\text{HR}}\cdot\frac{\rho V^2}{2} = 0.8\frac{0.0762\text{ m}}{0.326\text{ E - 3 m}}\cdot\frac{0.407\frac{\text{kg}}{\text{m}^3}\cdot\left(13.41\frac{\text{m}}{\text{s}}\right)^2}{2} = 6843\text{ Pa} = 0.99\text{ psi}$$

(f) This is 20 times that for the same size monolith, and explains why for automotive applications monoliths are almost the universal choice. For high pressure chemical processes, where small pressure drops are unimportant, most catalysts are pellets.

If we change the shape to make the bed frontal area twice as large and its length half as long, thus preserving the total volume, the velocity and length each decrease by a factor of 2, so that the pressure drop decreases by $2^3 = 8$. That helps, but it is not = to 20. The early automotive catalysts that used beads were designed to have the shape suggested here; the catalyst bed (in a porous metal cover) was shaped about like a textbook, with flow in the thin direction. Baffles directed that flow, with the thin direction being parallel to the road surface, for easy installation under the auto.

Chapter 8

8.1 I measured the first edition of this book. I assume the result will be the same for the second edition, when it is printed. (If they choose a different thickness paper, the result will be different). That book has 530 pages of text and is 19 mm thick, so the thickness of each page is

$$\text{thickness} = \frac{19\,\text{mm}}{\frac{530\ \text{pages}}{2}} = 0.071\frac{\text{mm}}{\text{page}} = 71\frac{\mu}{\text{page}}$$

The point of this problem is to give the students some feel for the size of a micron.

8.2** For any spherical particle $V = \frac{\pi}{6}D^3;\ A = \pi D^2;\ m = \rho V$

$$\frac{A}{m} = \frac{\pi D^2}{\rho V} = \frac{\pi D^2}{\rho \frac{\pi}{6} D^3} = \frac{6}{\rho D}$$

For a 1 cm diameter particle, $\frac{A}{m} = \frac{6}{62.3\frac{\text{lb}}{\text{ft}^3}\cdot 1\ \text{cm}} \cdot \frac{30.5\ \text{cm}}{\text{ft}} = 2.93\frac{\text{ft}^2}{\text{lb}}$

and for the other two sizes, if $D = 1\mu$, $\frac{A}{m} = 29{,}300\frac{\text{ft}^2}{\text{lb}} \approx 0.7\frac{\text{acres}}{\text{lb}}$;

and if $D = 0.01\mu$, $\frac{A}{m} = 2.93\ \text{E}6\frac{\text{ft}^2}{\text{lb}} \approx 70\frac{\text{acres}}{\text{lb}}$

One may also compute that for any size particle

$$\text{number} = \frac{\text{mass}}{\text{volume}} = \frac{m}{\rho\frac{\pi}{6}D^3};\quad \frac{\text{number}}{\text{mass}} = \frac{6}{\rho \pi D^3}$$

The number of particles for 1 cm diameter is

$$\frac{\text{number}}{\text{mass}} = \frac{6}{62.3\frac{\text{lb}}{\text{ft}^3}\cdot \pi \cdot (1\ \text{cm})^3}\cdot\left(\frac{30.5\cdot\text{cm}}{\text{ft}}\right)^3 = \frac{870}{\text{lb}}$$

For 1 μ it is 8.7 E4 particles per pound and for 0.01 μ it is 8.7 E20 particles per pound.

The point of this problem is to show that when the particles become very small, their surface area and (number/mass) both become very large.

8.3 (a) $m = \rho D^3;\quad D = \left(\frac{m}{\rho}\right)^{\frac{1}{3}} = \left(\frac{1\,\text{E-}6\ \text{g}}{1000\ \text{kg}/\text{m}^3}\right)^{\frac{1}{3}} = 1\,\text{E}-4\ \text{m} = 0.1\ \text{mm} = 100\ \mu$

By the same logic one can show that for a gram the answer is one cm, for a milligram it is one mm. When I was a student we had to use pan balances in quantitative analysis laboratory, and it was hard to get ± 1 mg on them. It is hard to believe that 1 mg is equivalent to a 1 mm cube of s.g. = 1 solid, but it is. Today's students have electronic balances, and miss the tedium and agony of pan balances.

(b) The above answer divided by $(2)^{1/3} = 79\ \mu$

After we cover the calculations I ask the students what solids have densities near 1 g/cm^3? The answer is ice, most organic compounds, lithium and sodium. Which solids have densities in the range of 2 to 3 g/cm^3? The answer is most rocks, aluminum and magnesium. Higher densities are almost exclusively metals or compounds high in metal content, e.g. metal ores.

I also ask which solids are cubical? Salt is, but few other solids are. Fly ash is often spherical, soot is lacy, many minerals are shaped like flakes or tablets, some (e.g. asbestos) are rod-like.

8.4** If we assume that then density of the coal and that of the ash are the same (only a mediocre assumption) then we can compute

$$\frac{m_{\text{ash}}}{m_{\text{particle}}} = 0.1 = \left(\frac{D_{\text{ash}}}{D_{\text{particle}}}\right)^3 = (0.1)^{\frac{1}{3}} = 0.46; \quad D_{\text{ash}} = 100\mu \cdot 0.46 = 46\mu$$

The smaller particles are partly formed by vaporization and subsequent condensation of volatile materials in the coal (as discussed in the text) and by the fact that the ash is not uniformly distributed in the coal. When the coal is ground to 100 μ particles, some will be mostly ash, some practically zero ash. However, to form a 1 μ ash particle on burning one would need a 100 μ coal particle which had 10^{-6} weight fraction = 0.0001 wt% ash. That is fairly unlikely, so the 1 μ particles must be mostly the products of vaporization and condensation. If one uses s.g. = 2, one computes an ash particle diameter of 37 μ.

8.5(a) The mass of a particle is

$$m_p = \rho V = \rho \frac{\pi}{6} D^3 = 2\frac{\text{g}}{\text{cm}^3} \cdot \frac{\pi}{6} \cdot (1\,\text{E}-6\ \text{cm})^3 = 1.05\,\text{E}-18\ \text{g} \approx 10^{-18}\,\text{g}$$

The mass of one carbon atom is

$$m_{\text{atom}} = \frac{12\dfrac{\text{g}}{\text{mol}}}{6.02\,\text{E}23\dfrac{\text{atoms}}{\text{mol}}} = 1.993\,\text{E}-24 \approx 2 \cdot 10^{-24}\frac{\text{g}}{\text{atom}}$$

so that the number of atoms per particle is $N = \dfrac{m_p}{m_a} = \dfrac{10^{-18}\,\text{g}}{2 \cdot 10^{-24}\dfrac{\text{g}}{\text{atom}}} = 50{,}000$ atoms

(b) If one had a particle with diameter 0.001 μ it would have a calculated mass of 50 atoms. That small an aggregation of atoms is not thermodynamically stable, and would be expected to either revert to the gas phase, or to grow rapidly by taking particles out of the gas. So if there are particles smaller than 0.01 μ they are unstable and have short lifetimes.

One of the reasons for assigning this problem is to make the students study Figure 8.1. Time spent learning all the useful information on that chart is time well spent.

8.6 (a) from a simple extrapolation on Figure 8.1 the settling velocity is of the order of 1 E6 cm/s 1 ft/yr.

(b) This is much less than typical wind velocities (see Chapter 5), for which reason we normally assume that gas molecules do not settle at all by gravity. That is an excellent engineering approximation, but they do settle, just at a negligible rate.

(c) The gas centrifuge method of separating uranium isotopes (as uranium hexafluoride) depends on this kind of settling. Gravity is replaced by centrifugal force, normally at thousands of times the acceleration of gravity. This process is used commercially in Europe to make nuclear reactor fuel and/or nuclear weapons-grade uranium 235.

Once in class a student was quite confused by this problem, because he had in mind the behavior of a cloud of some gas heavier than air, e.g. propane from a large propane release. Such clouds do move downward by gravity. However this problem asks for the behavior of a single molecule. I finally placated that student by referring to Stokes' law, which was written on the board for another problem. For the gas cloud which he had in mind the appropriate diameter was that of the cloud. For a single molecule, dispersed in the general atmosphere, it is the diameter of the molecule. That satisfied him.

One may also note that over geological time light gases like helium and hydrogen find their way to the top of the atmosphere by buoyancy, and sometimes escape off the top of the atmosphere. One can compute net upward velocities for them as done here; for thousands of years, they do have a significant net movement. I have also read that in deep natural gas wells, which are "shut in" for long periods of time, one finds a measurable difference in concentration of lighter and heavier gases with depth. The computation of the steady state value, which balances gravity with diffusion, is shown in some thermodynamics books.

8.7 (a) Small. See part (b)

(b) $$\text{Rate} = \frac{100\ \text{ton}}{\text{mile}^2 \cdot \text{month}} \cdot \frac{2000\ \text{lb}}{\text{ton}} \cdot \frac{\text{month}}{30\ \text{days}} \cdot \left(\frac{\text{mile}}{5280\ \text{ft}}\right)^2 = 2.4\ \text{E-4} \frac{\text{lb}}{\text{ft}^2\ \text{day}}$$

(c) $$\frac{\text{Depth}}{\text{time}} = \frac{\text{Rate}}{\rho} = 2.4\ \text{E-4} \frac{\text{lb}}{\text{ft}^2\ \text{day}} \cdot \frac{\text{ft}^2}{30\,\text{lb}} \cdot \frac{30\ \text{days}}{\text{month}} = 2.4\ \text{E-4} \frac{\text{ft}}{\text{month}} = 2.9\ \text{E-3} \frac{\text{inch}}{\text{month}}$$

In the early days of the air pollution effort, dustfall rates were regularly measured, because they are simple and cheap measurements to make. An uncovered petri dish lightly greased with Vaseline was exposed for a month and then weighed. Currently they are of little interest, because they mostly measure particles too large to be of human health consequence. Some European countries have regulations limiting the amount of heavy metals and/or fluorides in dustfall, to limit the ingestion of these materials by dairy cattle.

8.8** (a) From inside the back cover, the mass of a cubic foot of gas is 0.075 lb. The mass of particles is (100/7000) = 0.0143 lb, so the ratio is

$$\frac{\text{wt particles}}{\text{wt (gas + particles)}} = \frac{0.0143}{(0.075 + 0.0143)} = 0.160 = 16\%$$

(b) $$100\frac{\text{grains}}{\text{ft}^3} \cdot \frac{0.065 \text{ g}}{\text{grain}} \cdot \frac{35.29 \text{ ft}^3}{\text{m}^3} = 229\frac{\text{g}}{\text{m}^3}$$

(c) The mass of one particle is

$$m_p = \rho V = \rho \frac{\pi}{6} D^3 = 2\frac{\text{g}}{\text{cm}^3} \cdot \frac{\pi}{6} \cdot (1\text{ E-3 cm})^3 \cdot \frac{\text{lb}}{454 \text{ g}} = 2.3\text{ E-12 lbm}$$

so that the number of particles is

$$N = \frac{m_{total}}{m_p} = \frac{0.0143\dfrac{\text{lb}}{\text{ft}^3}}{2.3\text{ E-12}\dfrac{\text{lb}}{\text{particle}}} = 6.2\text{ E9 } \frac{\text{particles}}{\text{ft}^3}$$

(d) The logical guesses are a grain of sand or a grain of wheat. If we assume a specific gravity of 1 we can compute the diameter of a sphere which weights a grain, finding

$$D = \left(\frac{6}{\pi} \cdot \frac{m_p}{\rho}\right)^{1/3} = \left(\frac{6}{\pi} \cdot \frac{\text{lb}/7000}{62.3 \text{ lb}/\text{ft}^3}\right)^{1/3} = 0.0163 \text{ ft} = 0.196 \text{ in} = 4.99 \text{ mm}$$

This is much too large for sand, about right for wheat. Hence wheat is the most likely grain on which this measurement is based.

(e) In the US aspirin tablets are 5 grains. (Ibuprofen and acetomenaphin, which are more recent pain killers are labeled in mg!) The loadings in bullets are measured in grains. Older air conditioning charts all gave humidity in grains/per pound of dry air. Water hardness is often expressed in grains/gallon. A scruple is defined as 20 grains. Presumably someone who is unscrupulous doesn't give 20 grains! A grain is about 65 mg. In metric countries (and most US pharmaceuticals) the mg is the common small unit of mass.

8.9 From Figure 8.1 the settling velocity is 600 cm/s =13.4 mi/hr. The ratio of their horizontal to vertical travel will be (10 mi/hr)/(13.4 mi/hr) = 0.75. Thus they would be expected to fall within 0.75 stack heights of the base of the stack. These can be a local nuisance, but not an air pollution problem. Many communities have regulations requiring open trucks hauling sand to either cover it or wet it, to prevent it from blowing off the truck. This is for local nuisance prevention, not air pollution control.

8.10 From Fig 8.7 we can estimate the diameter as 2.2 μ. From Stokes' law (on which that part of Fig. 8.7 is based)

$$D = \sqrt{\frac{18\mu V}{9\rho_p g}} = \sqrt{\frac{18 \cdot \left(1.8\text{ E-}5\frac{\text{kg}}{\text{m}\cdot\text{s}}\right)\cdot\left(0.001\frac{\text{ft}}{\text{s}}\right)\cdot\frac{\text{m}}{3.281\text{ ft}}}{9\cdot\left(2000\frac{\text{kg}}{\text{m}^3}\right)\cdot\left(9.81\frac{\text{m}}{\text{s}^2}\right)}}$$

$$= 2.24\text{ E-}6\text{ m} = 2.24\ \mu$$

8.11** The density of the sphere is its mass divided by its volume

$$\rho = \frac{m}{V} = \frac{\rho_{\text{solid}}\frac{\pi}{6}\left(D_o^3 - D_i^3\right)}{\frac{\pi}{6}D_o^3} = \rho_{\text{solid}}\left(1 - \left(\frac{D_i}{D_o}\right)^3\right) = 2\frac{\text{g}}{\text{cm}^3}\left(1 - 0.98^3\right) = 0.1176\frac{\text{g}}{\text{cm}^3}$$

We can then substitute into Stokes' law and find the terminal settling velocity as

$$V = \frac{gD^2\rho_p}{18\mu} = \frac{\left(9.81\frac{\text{m}}{\text{s}^2}\right)\cdot(1\text{ E-}5\text{ m})^2\left(117.6\frac{\text{kg}}{\text{m}^3}\right)}{(18)\cdot\left(1.8\text{ E-}5\frac{\text{kg}}{\text{m}\cdot\text{s}}\right)} = 3.56\text{ E-}4\frac{\text{m}}{\text{s}} = 0.0356\frac{\text{cm}}{\text{s}}$$

Such hollow particles are regularly found in coal fly ash. The mechanism is presumably that in the furnace small fragments of molten ash collect on the surface of molten drops of coal, and agglomerate to form a solid crust. There are startling SEM photos of these particles in the literature. Some have one sphere within another, suggesting that perhaps the crust hardened, and then the drop inside formed a second crust, etc.

Students sometimes ask about the mass of the air inside the sphere. We ignored that here, as we did the buoyancy term in the common version of Stokes' law. If one takes one into account one should take the other into account, and the effect will cancel.

8.12 (a) For $\mathcal{R}_P = 0.3$, Stokes law gives $C_d = 80$, so Eq. (8.7) gives

$$C_D = 80\cdot\left(1 + 0.14(0.3)^{0.7}\right) = 80\cdot(1.060) = 84.8$$

(b) This is 1.06 times the Stokes' law value, so Eq. (8.7) gives a value 6% larger than Stokes' law.

This function does not exactly reproduce Stokes' law at low Reynolds numbers. This should help students see that this is an empirical data fit, which is probably good enough for practical work but not perfect. (How often will we really know the particle diameter to ± 6%?)

8.13 (a) Combining all three and solving for V^2 we find

$$V^2 = \left[\frac{\frac{4}{3} D \rho_P g}{\rho_f \cdot \left(\frac{24}{\mathcal{R}_P} (1 + 0.14 \mathcal{R}_P^{0.7}) \right)} \right] = \frac{\frac{4}{3} D \rho_P g \cdot \left\{ \frac{D V \rho_f}{\mu} \right\}}{24 \rho_f \cdot \left\{ 1 + 0.14 \cdot \left(\frac{D V \rho_f}{\mu} \right)^{0.7} \right\}}$$

Then clearing fractions we find

$$V + V^{1.7} (0.14) \left(\frac{D V \rho_f}{\mu} \right)^{0.7} - \frac{D^2 g \rho_P}{18 \mu} = 0$$

(b)

$$0.14 \cdot \left(\frac{D V \rho_f}{\mu} \right)^{0.7} = 0.14 \cdot \left(\frac{200 \text{ E - 6 m} \cdot 1.20 \frac{\text{kg}}{\text{m}^3}}{1.8 \text{ E - 5} \frac{\text{kg}}{\text{m} \cdot \text{s}}} \right) = 0.8582 \left(\frac{1}{\text{m/s}} \right)^{0.7}$$

and

$$\frac{D^2 g \rho_P}{18 \mu} = \frac{(200 \text{ E - 6 m}) \cdot \left(2000 \frac{\text{kg}}{\text{m}^3} \right) \cdot \left(9.81 \frac{\text{m}}{\text{s}^2} \right)}{18 \cdot \left(1.8 \text{ E - 5} \frac{\text{kg}}{\text{m} \cdot \text{s}} \right)} = 2.422 \frac{\text{m}}{\text{s}}$$

(c) I did the solve on an HP 15C, finding that result.

The problem of replacing the trial and error solution for the settling velocity for particles too large for Stokes' law to apply is explored in some detail in McCabe. W.L. and J.C. Smith, *Unit Operations of Chemical Engineering* 3e, McGraw-Hill , N.Y. 1976, p 154 et. seq. The solutions shown there (and here) are useful if one is programming this type of calculation on a computer, but not particularly useful for hand solutions.

This problem and the solutions cited above do not really change this from being a trial-and-error solution. The resulting equations are transcendental, and can only be solved by trial and error. Our hand calculators have swift and efficient trial-and-error routines built into them so that although it does not appear to us that they are solving by trial-and-error (e.g. Newton's method) but they are.

8.14 From inside the back cover of the book ρ = 4.16 E-5 mol/cm^3 so that

$$\lambda = \frac{1}{\sqrt{2} \cdot \pi \cdot (6.02 \text{ E23 / mol}) \cdot \left(4.16 \text{ E - 5} \frac{\text{mol}}{\text{cm}^3} \right) \cdot (3.48 \text{ E - 10 m})} \cdot \left(\frac{\text{m}}{100\text{cm}} \right)^2$$

$$= 7.42 \text{ E - 6 cm} = 0.07 \ \mu$$

which is the value used in Example 8.4.

8.15 (a) It rotates in the direction of the mirrors, i.e. the mirrors go forward, the black sides go backward. It does this because in the vacuum normally achievable with simple vacuum pumps there are enough air molecules that some of them adhere to the surfaces of the plates. The black side is hotter than the silver side, due to absorption of incoming light. The air molecules that become absorbed and then desorb on the black side leave at a higher temperature and hence at a higher momentum than those on the mirror side, and hence they impart more reaction force, etc.

(b) In a perfect vacuum it would rotate the opposite direction. In that case there would be no adsorbed air molecules. The photons which hit the black side would be absorbed. The photons which hit the mirror side would be reflected, and thus transfer twice as much momentum to the solid as those that are absorbed on the black side. I do not know if this has been demonstrated experimentally; it would require a very good vacuum.

c) The Cunningham correction factor is caused by the adsorption and re-emission of air (or gas) molecules, as described in (a).

This problem has the merit that one can bring a Crookes radiometer to class and show that it works the way described above. Crookes radiometers spin fast in direct sunlight, slowly when held up to a fluorescent lamp, not at all in ordinary room illumination. I hope that helps the students understand the Cunningham correction factor better.

The "technical folklore" teaches that Crookes expected it to move the way shown in (b) and was surprised when it moved the other way. I have no good evidence that this part of the folklore is true.

8.16**(a) See Example 8.2, page 173. The answer here is twice the value there because the density is twice as large, or 1.99 · 2 = 3.98 E-4 ft/s = 1.2 E-4 m/s = 0.012 cm/s.

(b) See Example 8.4. $1 + A\frac{\lambda}{D} = 1 + 1.728 \cdot \left(\frac{0.07\ \mu}{1\ \mu} \right) = 1.21$

$$V = 3.98\ \text{E}-4 \frac{\text{ft}}{\text{s}} \cdot 1.21 = 4.46\ \text{E}-4 \frac{\text{ft}}{\text{s}} = 0.01359 \frac{\text{cm}}{\text{s}} \approx 0.014 \frac{\text{cm}}{\text{s}}$$

This is assigned as a vehicle for the discussion of the Cunningham correction factor. I reiterate to the students that the CCF is a Taylor series of which we usually only use the first term. For particles smaller than about 0.4 μ in diameter one should find and use the forms with the higher powers of (λ/D). See ref [1].

8.17 (a) The diameter of the baseball is 0.074 m, and its density 0.694 g/cm^3. If it moves 69 diameters that corresponds to 5.08 m. Substituting these values in the Eq. 8.12 and solving for the viscosity we find

$$\mu = \frac{VD^2\rho C}{18x_{ss}} = \frac{\left(10\frac{\text{m}}{\text{s}}\right) \cdot (0.074\ \text{m})^2 \cdot \left(694\frac{\text{kg}}{\text{m}^3}\right) \cdot (1)}{(18) \cdot (5.08\ \text{m})} = 0.416\ \text{Pa} \cdot \text{s}$$

$$= 416\ \text{cp} = 4.16\ \text{poise}$$

which is 416 times the viscosity of room temperature water, comparable to the viscosity of cold lube oil.

(b) The particle Reynolds number is

$$\mathcal{R}_P = \frac{DV\rho}{\mu} = \frac{(0.074 \text{ m})\cdot\left(10\dfrac{\text{m}}{\text{s}}\right)\cdot\left(900\dfrac{\text{kg}}{\text{m}^3}\right)}{0.416 \text{ Pa}\cdot\text{s}} = 1600$$

From Equation 8.7 we can estimate that

$$\left(\frac{C_D}{C_{D_{\text{stokes}}}}\right)_{\text{initial}} = \left(1 + 0.14\mathcal{R}_P^{\;0.7}\right) = 25.5$$

and that the stopping distance would be roughly (1/25) of the above value or 0.2 m 0.7 ft. The fact that the particle Reynolds number is this high shows that the assumption of stokes drag force is inappropriate. Here if we did the experiment, with a 416 cp fluid, we would probably have some splashing, but the predicted travel distance of 0.7 ft would probably be observed.

8.18**(a) Going back to Eq. 8.10, we separate variables and integrate, finding

$$\int \frac{dV}{V} = \frac{-18\mu}{D^2\rho C}dt; \quad \ln\frac{V}{V_o} = \frac{-18\mu}{D^2\rho C}\Delta t; \quad \Delta t = \frac{D^2\rho C}{18\mu}\ln\frac{V_o}{V}$$

then setting $V = 0$, we find that the required time is infinite.

(b) Setting $(V_o/V) = 100$, we find

$$\Delta t = \frac{(1\text{ E-6 m})^2\cdot\left(2000\dfrac{\text{kg}}{\text{m}^3}\right)\cdot 1.12}{18\cdot\left(1.8\text{ E-5}\dfrac{\text{kg}}{\text{m}\cdot\text{s}}\right)}\ln 100 = 3.2\text{ E-5 s}$$

(c)Initially

$$\mathcal{R}_p = \frac{DV\rho_{\text{air}}}{\mu} = \frac{(1\text{ E-6 m})\cdot\left(10\dfrac{\text{m}}{\text{s}}\right)\cdot\left(1.2\dfrac{\text{kg}}{\text{m}^3}\right)}{1.8\text{ E-5}\dfrac{\text{kg}}{\text{m}\cdot\text{s}}} = 0.67$$

(d) Initially

$$\frac{C_D}{C_{D\text{stokes}}} = \left(1 + 0.14\mathcal{R}_p^{\;0.7}\right) = \left(1 + 0.14\cdot(0.67)^{0.7}\right) = 1.105$$

Finally

$$\frac{C_D}{C_{D\text{stokes}}} = 1$$

So on the average the drag coefficient is 1.05 times the assumed Stokes law value, so that the stopping distance should be (1/1.05) 0.95 times the value calculated in Example 8.5.

8.19 The sketches for parts (a) and (b) are shown below.

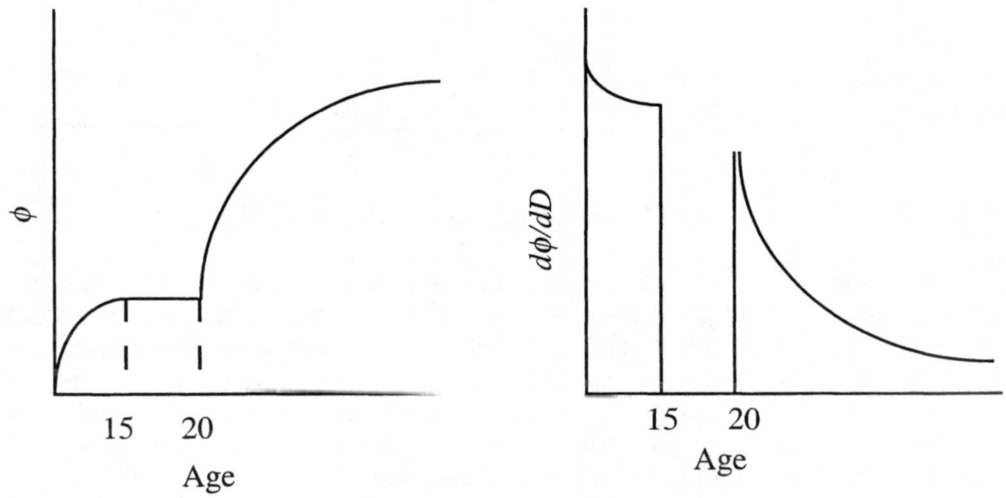

This is a simplified version of the effect of major wars, epidemics, etc. I use it as the first distribution problem because it has a very high intuitive content.

8.20 (a) The three sketches are shown below

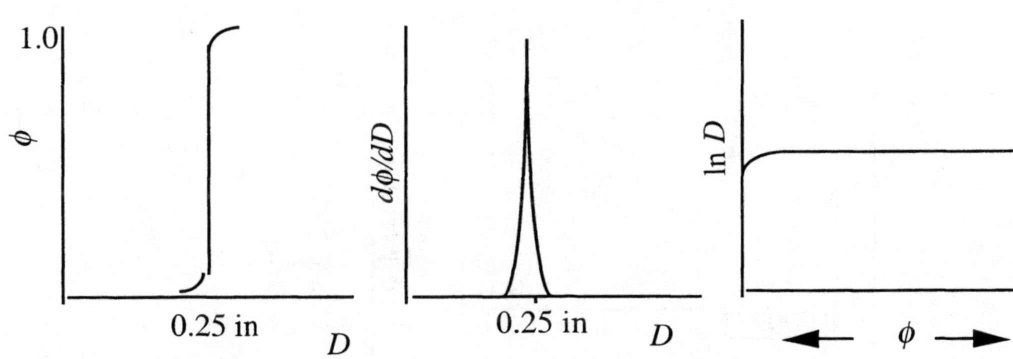

(b) The three sketched are shown below

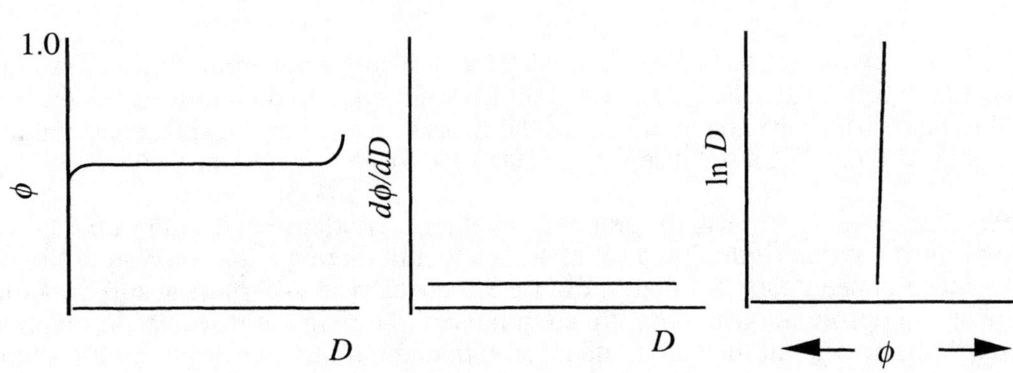

(c) The three sketches are shown below

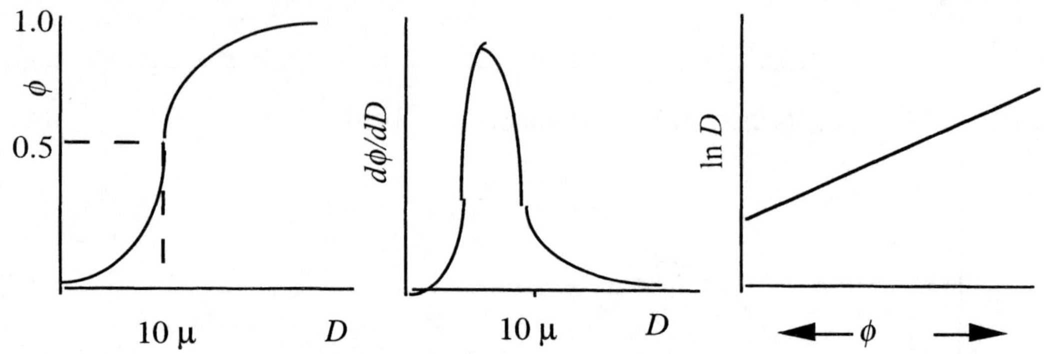

These are the closest I can come to Gaussian curves on MacDraw II.

(d) Part (a) corresponds to ball bearings balls. I hope I have the permitted sigma right. Part (b) corresponds to all the spherical particles in the universe, from protons to galaxies. When I first gave the problem I used sigma = 100, and a student showed me that with that high a sigma we would have negligible chance of finding either protons or galaxies. This example was chosen simply to show the students that a large value of sigma leads to a very broad distribution. Part (c) is a typical distribution of particles in a flue gas. I originally thought that in (c) the maximum in the $d\phi / dD$ curve would be at 10 μ, but the students ran it on their spreadsheets and found that it is greater at a smaller diameter. Here we are plotting a log-normal distribution on normal coordinates, which causes this result. If we plotted $d\phi / d\ln D$ vs ln D, then the maximum would occur at 10 μ.

8.21 See Example 8.9. The sketches for parts (a) and (b) are shown below

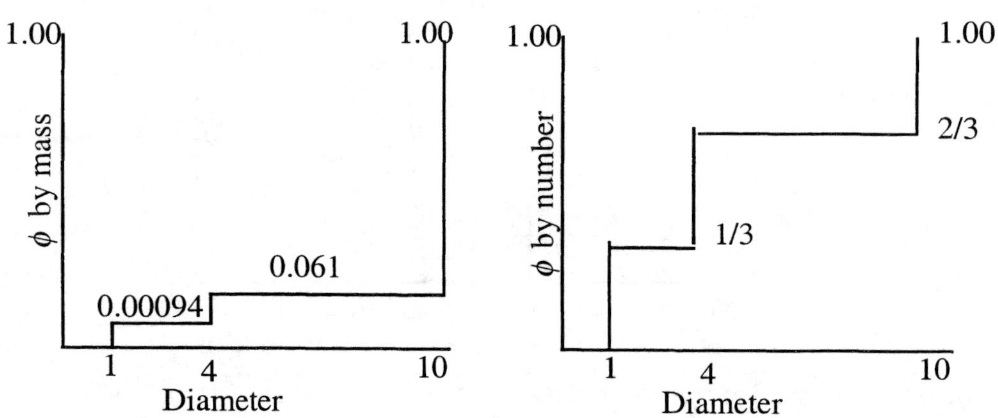

After you have discussed this you can ask the students what a plot of $d\phi/dD$ vs D would look like? It has vertical spikes at 1, 4 and 10 microns. In the number plot each of the spikes has width zero and area 1/3. In the mass plot each of the spikes has width zero, but the areas are (1/1065), (64/1065) and (1000/1065).

After that discussion, I ask the students whether such steps exist in Figure 8.8. Eventually they figure out that figure has a set of steps too, but there are 248 million of them, so that one does not see them. Similarly, Figure 8.9 consists of 248 million spikes, with the curve shown an appropriate average. We are not used to looking at distribution functions for groups this small, but they are quite informative about the idea of distribution functions.

8.22** (a) $z = \dfrac{\ln\left(\dfrac{D}{D_m}\right)}{\sigma} = \dfrac{\ln\left(\dfrac{1}{5}\right)}{0.8} = -2.01$

From Table 8.3 for $z = +\ 2.01$, $\phi = 0.9772$, so that $1\text{-}\phi = 0.0228 = 2.28\%$

(b) Using Eq. 8.32 and following Example 8.10

$$D_{\text{m number}} = D_{\text{m weight}} \exp(0-3)\sigma^2 = 5\,\mu \exp(-3)\cdot(0.8)^2 = 0.773\,\mu$$

$$z = \frac{\ln\dfrac{1}{0.733}}{0.8} = 0.388; \qquad \phi \approx 0.65$$

65% by number of the particles have $D \leq 1\,\mu$

This illustrates this convenient property of the log normal distribution. As far as I know that is not a property of any other distribution function. One can illustrate the idea by a rabbit and elephant stew. If one uses one rabbit and one elephant, then it is 50% rabbit by number, but far from 50% rabbit by mass.

8.23 For b =1 we would have found the distribution by length, and for $b = 2$ the distribution by surface area. That distribution is referred to in the spraying literature as the Sauter distribution, or more often, the mean diameter by area is referred to as the "Sauter Mean Diameter". Figure 8.11 shows the distribution by surface area of atmospheric particles by area, because that makes a better plot than one by mass or number.

8.24(a) No. Here

$$z_{100} = \frac{100-10}{\sigma} = \frac{90}{\sigma}; \quad z_1 = \frac{1-10}{\sigma} = \frac{-9}{\sigma}$$

for a normal distribution, and the assumed equal frequency of occurrence, these should have equal absolute values.

(b) Yes. Here

$$z_{100} = \frac{\ln\dfrac{100}{10}}{\sigma} = \frac{2.303}{\sigma}; \quad z_1 = \frac{\ln\dfrac{1}{10}}{\sigma} = \frac{-2.303}{\sigma}$$

these have equal absolute values.

(c) The two sketches are shown below.

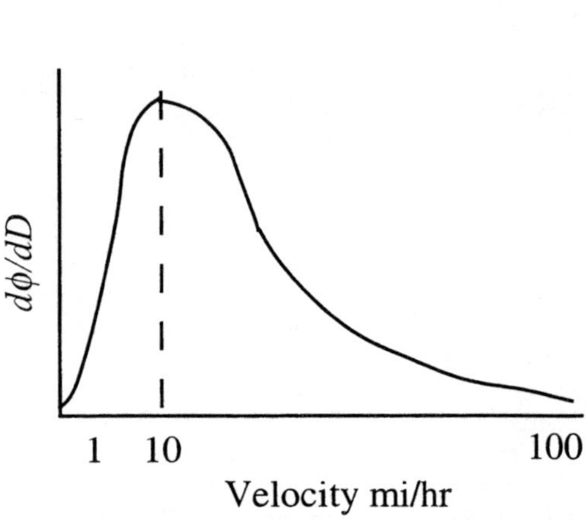

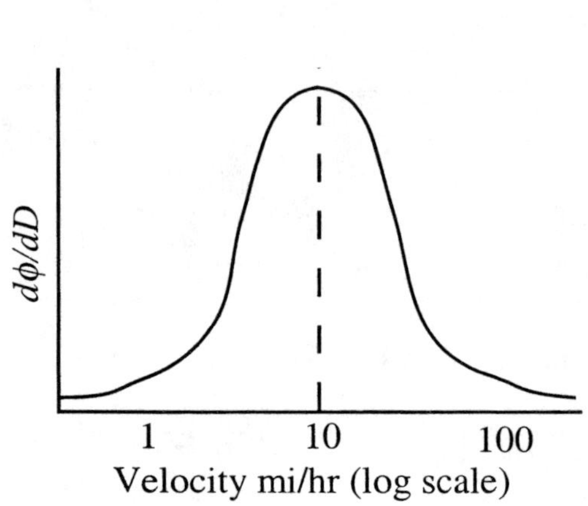

As far as I know almost all weather data and data on things like annual stream flow are better represented by the log-normal distribution than the normal distribution.

8.25 The easiest approach is to plot these data on normal and log-normal papers. The values form a pretty good straight line on log-normal paper, a much poorer one on normal paper.

8.26** Here one can simply extrapolate, or one can be clever and see the symmetry property ($D_{50}/D_{10} = D_{90}/D_{50}$ etc.) and use that to get out of the extrapolation by reading the values for 90% and 99%. But the best approach is to learn use the table and thus calculate the values.

$$z_{10\%} = -1.28, \quad z_{1\%} = -2.32 \text{ (Both from Tab. 8.3)}$$
$$D_{10\%} = D_m \exp(z\sigma) = 20\ \mu \exp(-1.28 \cdot 1.25) = 4.04\ \mu$$
$$D_{1\%} = 20\mu \exp(-2.32 \cdot 1.25) = 1.1\ \mu$$

The point of all of this is to learn how to replace experimental data with simple equations, and then use those. Problem 8.29 shows the equation which I believe most computers use to compute the table when they need it.

8.27 (a) First we observe from Fig. 8.10 that 10 μ corresponds to 30% by weight of the sample. Then we draw the first two plots, and draw on them the line representing a perfect split at 10 μ. Then we observe that this hypothetical collector did not change the relations between sizes in the size range below 10 μ, so that the shape of the curves for the remaining particle set is the same as it was for that part of the incoming particle set but rescaled. The curves are those sketched below, in which the shaded area remains and the vertical scale is multiplied by a factor of (1/0.3).

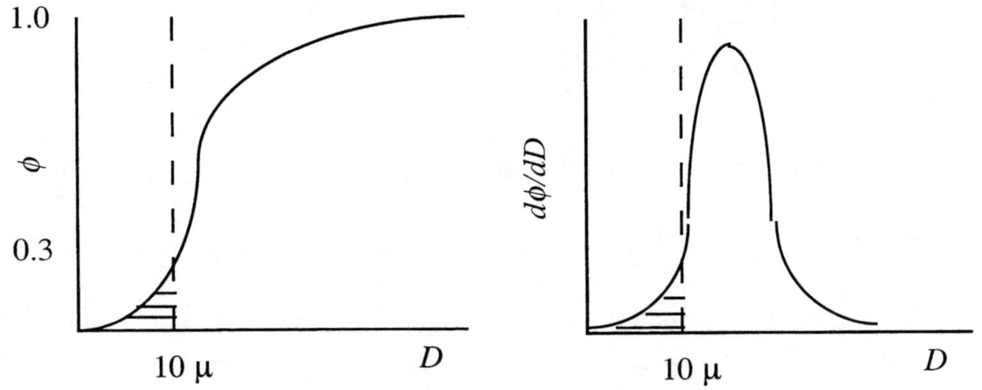

It comes as a shock to students that one can simply divide and rescale to get the distribution. That is only true for a collector which has a perfectly sharp collection cutoff, which no real ones do. Eventually most understand this. I then ask the students whether the particle set which passes through is log-normally distributed. About half say "yes". Then I ask them how it would look on a plot like Fig 8.10. It is clear that it must become asymptotic to 10 μ as the cumulative fraction approaches 1, so that the curve must become horizontal. Thus it cannot be a straight line on that coordinate system, and hence cannot obey the log normal distribution.

(b) The mass mean diameter of the particles that pass through corresponds to the 15% by weight point of the original distribution. (This puzzles students, but eventually they get it.) So

$$z_{15\%} = -1.05$$

$$D_{\text{mass mean}} = D_m \exp(z\sigma) = 20\mu \exp(-1.05 \cdot 1.25) = 5.38\ \mu$$

We calculate the number mean of the original distribution using Eq. 8.32

$$D_{\text{mean number}} = \frac{D_{\text{mean mass}}}{\exp(3\sigma^2)} = \frac{20\ \mu}{\exp(3 \cdot 1.25^2)} = 0.184\ \mu$$

Then, for the distribution by number we compute that 10 μ corresponds to

$$z = \frac{\ln\left(\dfrac{10}{0.184}\right)}{1.25} = 3.196$$

From Tab. 8.3 we see that this corresponds to ϕ = 0.9993. Thus although the collector collects 70% of the weight of the particles, it collects only 0.07% by number of the particles. The number mean diameter of the particles that pass through corresponds to ϕ = (0.9993/2) = 0.4965 0.5, so that the mean diameter by number is practically that of the original distribution, or 0.184 μ.

This latter calculation helps persuade the students that almost all mass is in the large particles and almost all the number is in the small particles.

8.28 For 99.9%, $\quad z = 3.1; \quad D = D_m \exp(z\sigma) = 10\ \mu \cdot \exp(3.1 \cdot 1) = 221.9\ \mu$

For 0.01 %

$$1-\phi=(1-.0001)=0.9999; \quad z=-3.7$$
$$D=10\ \mu\exp(-3.7\cdot 1)=0.25\ \mu$$

8.29 I did these calculations on a spreadsheet

(a) I find that for (1-ϕ) = (1- 0.9772) = 0.0228, the calculated value of z = 1.99966 2.0.

(b) For (1-ϕ)= 1 E-9, the calculated value of z = 5.99987 6. This is similar to the values at the end of Ex. 8.8.

I got this formula from a book on algebraic approximations for use in computers [16]. I assume that statistical packages on computers use this formula or a similar one.

8.30 This problem and the following two are quite unrealistic physically, but show the students how one would perform the equivalent of Examples 7.5 and 7.6 if one had suitable analytic function representations of the distribution function and of the collection efficiency as a function of particle size. In Example 9.6 the same process is carried out numerically.**

Starting with Equation 7.10 we substitute and integrate. As a preliminary step we recognize that the efficiency is given by $\eta=\left(\frac{D}{D_{\max}}\right)^2; \quad p=1-\eta$

$$p_{\text{overall}}=\int_0^{D_{\max}}\left[1-\left(\frac{D}{D_{\max}}\right)^2\right]\frac{1}{D_{\max}}dD=\left(\frac{D}{D_{\max}}-\frac{1}{3}\frac{D^3}{D_{\max}{}^3}\right)\Bigg|_0^{D_{\max}}$$
$$=\left(1-\frac{1}{3}\right)=0.667; \quad \eta=1-0.667=0.333$$

8.31 See the solution to Prob. 8.30. Again we could substitute into Eq. 7.10 and integrate. However, because of the form of the functions chosen, it is easier to work in efficiencies, rather than in penetrations (if we are doing the problem by hand). So

$$\eta_{\text{overall}}=\int\eta(D)\frac{d\phi}{dD}dD=\int_0^{10}aD^2bDdD+\int_{10}^{20}aD^2b(e-D)Dd$$
$$=\frac{abD^4}{4}\Bigg|_0^{10}+abe\frac{D^3}{3}\Bigg|_{10}^{20}-\frac{abD^4}{4}\Bigg|_{10}^{20}$$
$$=\left(\frac{0.0025}{\mu^2}\right)\cdot\left(\frac{0.01}{\mu^2}\right)\cdot\left\{\frac{10^4\mu^4}{4}+\frac{20\cdot 8\cdot 10^3}{3}\mu^4-\frac{10^4\mu^4}{3}-\frac{20^4\mu^4}{4}+\frac{10^4}{4}\mu^4\right\}$$
$$=0.291=29.1\%$$

8.32** See the solution to Prob. 8.31. Again we work in efficiencies. For the given values of k_1 and k_2 there are no particles large enough for the $\eta = 1.0$ part of the collection efficiency relation to come into play, so that

$$\eta_{\text{overall}} = \int \eta(D)\frac{d\phi}{dD}dD; \quad \frac{d\phi}{dD} = 2k_1 D$$

$$= \int k_2 D \cdot 2k_1 D dD = 2k_1 k_2 \left.\frac{D^3}{3}\right|_0^{\frac{1}{\sqrt{k_1}}} = \frac{2}{3}\frac{k_2}{\sqrt{k_1}}$$

$$= \frac{2}{3} \cdot \frac{1/\mu}{\sqrt{0.01/\mu^2}} = \frac{2}{3} = 66.7\%$$

8.33 This problem is very similar to Problem 8.27, and the discussion there is applicable to it.

(a)

$$z = \frac{\ln\frac{D}{D_m}}{\sigma} = \frac{\ln\frac{5}{10}}{1.5} = -0.462$$

$$\phi = 1 - 0.67 = 33.$$

So 67% are caught and 33% pass through.

(b) The three sketches are shown below. The explanation is in the solution to Prob. 8.27.

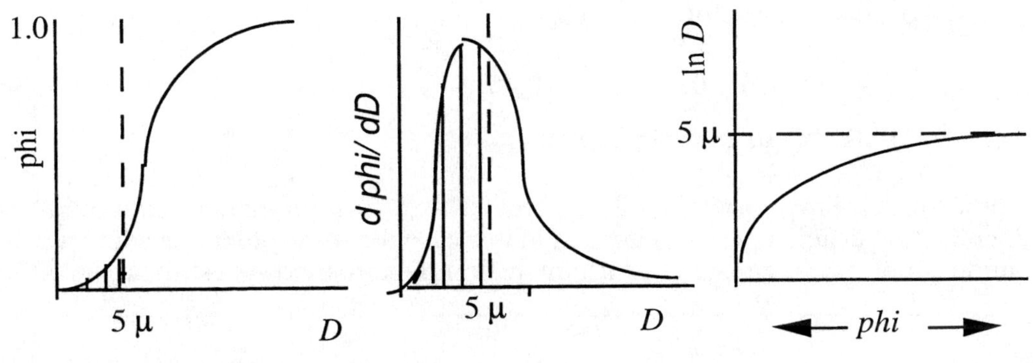

(c) If 33 % pass through, then the 50% point of those that pass through is at the 16.5% point of the original distribution and

$$1 - \phi = 1 - 0.165 = 0.835; \ z = 0.95$$

$$D = D_m \exp(z\sigma) = 10\mu \exp(-0.95 \times 1.5) = 2.4\mu$$

If 67% are caught, then the 50% point of those that are caught is 1- 0.67/2 = 0.665 of the original distribution and $z = 0.427$

$$D = 10\mu \exp(0.427 \cdot 1.5) = 19\mu$$

8.34 We make up the following table

Diameter, microns	$z = \dfrac{\ln(D/5)}{1.5}$	ϕ from Tab. 8.3
4	-0.149	1-0.56 = 0.44
2	-0.610	1-0.73 = 0.27

$$\eta = 0.56 + 0.5 \cdot (0.44 - 0.27) = 0.56 + 0.085 = 0.645 \approx 65\%$$

8.35** (a) First we compute that for 40 μ particles, and 10 μ particles

$$z_{40} = \frac{\ln \dfrac{40}{10}}{1.5} = 0.924; \quad \phi = 0.82; \quad z_{10} = \frac{\ln \dfrac{10}{10}}{1.5} = 0; \quad \phi = 0.5$$

Then we make up the following table

Diameter, microns	$\Delta\phi$	η	p	$\sum p\Delta\phi$
0-10	0.5	0	1.00	0.50
10-40	0.32	0.5	0.5	0.16
40+	0.18	1.00	0	0

The overall penetration is 0.50 + 0.16 +0 = 0.66 = 66%, and fraction collected = 34%.

(b) 66% pass through uncollected, so the 50% point of that group corresponds to the 33% point of the original distribution. For it

$$\phi = 0.33; \quad (1-\phi) = 0.67; \quad z = -0.45$$

$$D = 10\ \mu \exp(-0.45 \cdot 1.5) = 5.09\ \mu$$

This procedure is only correct if, of the particles that pass through, the first 50% have the same collection efficiency, either zero as in this case, or some other constant value. If that condition is not met, then one must use more complex methods to estimate the 50% point.

8.36 (a) Here we choose as our basis that mass of particles which has 1 m^2 of surface area. Then each of the three groups will have 1/3 m^2 of surface area.

$$A_i = N_i \pi D_i^2; \quad M_i = N_i \frac{\pi}{6} D_i^3 \rho; \quad \text{eliminating } N_i$$

$$M_i = \frac{\pi}{6} D_i^3 \rho \cdot \frac{A_i}{\pi D_i^2} = \frac{A_i D_i \rho}{6}$$

$$M_{\text{small}} = \left(\frac{1}{3}\text{m}^2\right) \cdot (0.02\ \text{E-6 m}) \cdot \left(2000 \frac{\text{kg}}{\text{m}^3}\right) \cdot \frac{1}{6} = 2.22\ \text{E-6 kg}$$

and correspondingly the masses of the medium and large particles are 33.3 E-6 and 1111 E-6 kg. The total is 1146.5 E-6 kg.

The fraction of the total mass in the smallest particles is $\frac{2.22}{1146.5} = 1.93\text{ E} - 3$, in the medium particles is $\frac{33.3}{1146.5} = 0.029$ and in the largest particles is $\frac{1111}{1146.5} = 0.969$

(b) $N = \frac{A_i}{\pi} D_i^2$

The number of small particles is $N_{\text{small}} = \frac{\left(\frac{1}{3}\text{m}^2\right)}{\pi \cdot (0.02\text{ E - 6 m})^2} = 2.62\text{ E15}$

correspondingly, N_{medium} = 1.179 E12, N_{large} =1.05 E10. The sum is 1.624 E 15

The fraction of the total number in the smallest particles is $\frac{2.62}{2.624} = 0.9995$

in the medium particles is $\frac{0.00018}{2.624} = 0.00045$

and in the largest particles is $\frac{1.05 \times 10^{10}}{2.624 \times 10^{15}} = 4 \times 10^{-6}$

Chapter 9

9.1 We can calculate the distance x at which a particle which enters at any z reaches the floor of the settler by

$$\frac{dz}{dt} = -V_t; \quad \frac{dx}{dt} = V = \frac{6}{H^2} V_{avg} \cdot (zH - z^2); \; dx = \frac{\frac{6}{H^2} V_{avg} \cdot (zH - z^2)}{V_t} \cdot (-dz)$$

$$x = \left[\frac{6}{H^2} \cdot \frac{V_{avg}}{V_t} \left(\frac{z^3}{3} - \frac{z^2}{3} \right) \right]_{z_0}^{0} = \frac{V_{avg}}{V_t} \cdot \frac{z_0^{\,2}}{H^2} \cdot (3H - 2z_0)$$

If we take the derivative of this with respect to z, set it equal to zero and solve for z we find $z = H$ which is the not-surprising result that the particle which enters at the top of the chamber falls farthest from the entrance. Then substituting $z = H$ in the above, we find

$$\frac{dx}{dz_0} = \frac{V_{avg}}{V_t H^2} \cdot \left(6Hz_0 - 6z_0^{\,2} \right); \quad = 0 \;\; \text{when } z_0 = H$$

which is the same result as for block flow! This is a somewhat artificial problem, but it shows that the velocity distribution is unimportant for any single-chamber gravity settler. However if the settler were internally divided by horizontal collector plates, as some are, then the required length would be set by the chamber which had the highest velocity, in this case the center chamber. For laminar flow the center velocity is twice the average velocity.

For the very small gravity settlers used for particulate sampling (called *horizontal elutriators*), the flow is surely laminar; this would be the logical description of their behavior.

9.2 We start with Equations 9.4 and 9.11, and substitute the equation for the terminal settling velocity from Example 8.4, finding $V_t = \sqrt{\frac{8}{6} \cdot \frac{\rho_p D g}{\rho_g C_D}}$

$$\eta_{block} = \frac{L}{HV_{avg}} \sqrt{\frac{4}{3} \cdot \frac{\rho_p D g}{\rho_g C_D}}; \quad \eta_{mixed} = 1 - \exp(\eta_{block})$$

From Table 8.1 it is clear that one can use Stokes' law up to a particle diameter of 70 μ with only a 10% error. Thus these modified equations would be appropriate if one were using a settling chamber to collect sand-size particles, but not for the particles of air pollution interest.

9.3 From example 8.2, V_t = 6.05 E-5 m/s. Then

$$V_{avg} = \frac{LV_t}{H\eta} = \frac{75 \text{ mm} \cdot 6.05 \text{ E-5 m/s}}{2 \text{ mm} \cdot 0.9} = 2.51 \text{ E-3} \frac{\text{mm}}{\text{s}}$$

One may check the Reynolds number, finding 0.3, so the flow is certainly laminar. The Cunningham correction factor is about 1.12, which would lead to multiplying the above velocity by 1.12.

The Industrial Hygiene licensing exam is closed-book multiple choice. A question about horizontal elutriators appears regularly. Apparently they were a specified standard method in England for many years.

9.4 Here there are two buoyant forces, one in the opposite direction to the centrifugal force, the other in the direction opposite to gravity. Both are equal to the force in the opposite direction (centrifugal or gravity) times (air density/particle density). Both are negligible in most gas-particle problems, but not in liquid-particle problems..

9.5 In example 9.2 the ratio of centrifugal/gravity was 111.8. In the first part of (a) the radius is reduced by a factor of 10, so the ratio is increased by a factor of 10 to 1118. In the second part of (a) the velocity is increased by a factor of 2, compared to the example, so that the ratio is increased by a factor of 4, to 447.

(b) At the equator

$$\frac{\text{centrifugal}}{\text{gravity}} = \frac{V_c^2}{rg} = \frac{(1000\ \text{mi / hr})^2}{4000\ \text{mi} \cdot 32.2\ \text{ft / s}^2} \cdot \frac{5280\ \text{ft}}{\text{mi}} \cdot \left(\frac{\text{hr}}{3600\ \text{s}}\right)^2 = 3.2\ \text{E - 3}$$

(c) To make this equal to one, V must increase by $\dfrac{1}{\sqrt{3.2\ \text{E - 3}}} = 17.8$

Ask the students where they have experienced strong centrifugal forces? Autos on sharp turns, bicycles on which one leans into turn, various rides at amusement parks. On the new coaster rides in which the rider is head down one can watch those with long hair and see that it always points outward, so that even at the top of the loop, the vector sum of gravity plus centrifugal is driving the riders into their seats.

9.6 Writing equation 9.18 twice, once for the cut diameter and once for any diameter, and taking the ratio, and then taking the square root of that, we find

$$\frac{\eta}{\eta_c} = \left(\frac{D}{D_c}\right)^2; \quad \frac{D}{D_c} = \sqrt{\frac{\eta}{0.5}}; \quad \text{for } \eta = 0.1, \ \frac{D}{D_c} = \sqrt{\frac{0.1}{0.5}} = 0.447$$

$$\text{for } \eta = 0.9, \ \frac{D}{D_c} = \sqrt{\frac{0.9}{0.5}} = 1.34$$

We rearrange equation 9.19 to

$$\eta - 1 = \exp\left(-\alpha D^2\right); \quad \text{where } \alpha = \frac{\pi N V_c \rho_p}{9 W_i \mu}$$

Then, writing it for the cut diameter we solve for α, finding $\alpha = \dfrac{0.693}{D_c^2}$

substituting this back into Equation 9.19, rearranging and taking the square root, we find

$$\ln(1-\eta) = -\,0.693 \cdot \left(\frac{D}{D_c}\right)^2$$

$$\frac{D}{D_c} = \sqrt{-\frac{\ln(1-\eta)}{0.693}}; \quad \text{for } \eta = 0.1 \quad \frac{D}{D_c} = 0.3899; \quad \text{for } \eta = 0.9 \quad \frac{D}{D_c} = 1.823$$

Equation 9.21 is rearranged and the square root taken, finding

$$\frac{D}{D_c} = \sqrt{\frac{\eta}{1-\eta}}; \quad \text{for } \eta = 0.1 \quad \frac{D}{D_c} = 0.333; \quad \text{for } \eta = 0.9 \quad \frac{D}{D_c} = 3$$

These are indeed the values used to make up Figure 9.6.

9.7 From equation 9.18 the cut diameter is proportional to 1/square root of inlet velocity, so that

$$\frac{D_{c_{\text{new}}}}{D_{c_{\text{old}}}} = \sqrt{\frac{V_{old}}{V_{new}}} = \sqrt{\frac{1}{1.25}} = 0.8944$$

$$D_{c_{\text{new}}} = 5\,\mu \cdot 0.8944 = 4.47\mu$$

9.8 (a) Writing Equation 9.18 twice, once for the cut diameter, once for an arbitrary particle diameter, taking the ratio and simplifying, we find

$$\eta = 0.5 \cdot \left(\frac{D}{D_c}\right)^2; \quad p_1 = 1 - 0.5 \cdot \left(\frac{D}{D_c}\right)^2; \quad p_3 = (p_1)^3 = \left(1 - 0.5 \cdot \left(\frac{D}{D_c}\right)^2\right)^3$$

If we set $p_3 = 0.5$ we find the cut diameter for three in series. Taking the cube root and simplifying $\dfrac{D_{c_3}}{D_{c_1}} = \sqrt{\dfrac{1 - 0.5^{\frac{1}{3}}}{0.5}} = 0.642$

(b) Writing the equation for the cut diameter twice (for old and new) and taking the ratio we find $\dfrac{D_{c_{\text{new}}}}{D_{c_{\text{old}}}} = \sqrt{\dfrac{N_{\text{old}}}{N_{\text{new}}}} = \sqrt{\dfrac{1}{3}} = 0.577$

The plot is shown at the right. At low values of D/D_{cut} the combination behaves like one cyclone with $N = 15$. At high values it behaves like one cyclone with $N = 5$. Compare this to Figure 9.2.

This is probably a good place to remind the students that N is not determined theoretically; the value of $N = 5$ is chosen because it fits the observed behavior of many cyclones.

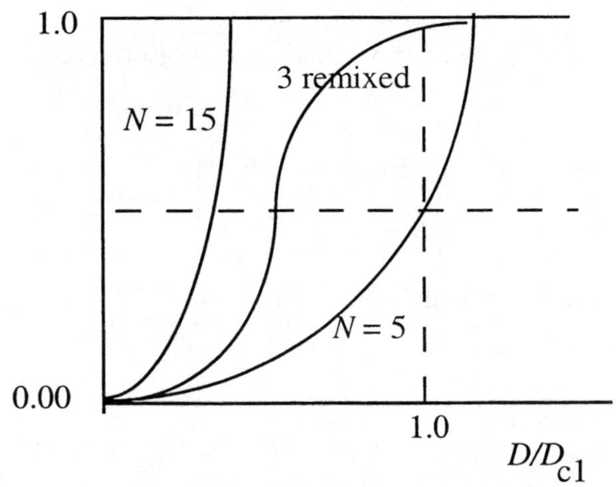

The only industrial example I know of in which cyclones are regularly connected in series is on the exhaust gas from catalytic cracker regenerators. There the particles are 200 μ in diameter, and the particle loading in the gas is very large. The two in series arrangement is probably chosen because of the high particle loading, rather than to lower the cut diameter.

9.9 $$W_i = \frac{2\pi D^2 N V_c \rho_p}{9\mu}$$

$$= \frac{2\pi \cdot (1\text{ E-6 m})^2 \cdot 5 \cdot 50\frac{\text{ft}}{\text{s}} \cdot \frac{2000\text{ kg}}{\text{m}^3}}{(9)\cdot\left(1.8\text{ E-5}\frac{\text{kg}}{\text{m}\cdot\text{s}}\right)} \cdot \frac{\text{m}}{3.281\text{ft}} = 5.91\text{mm} = 0.23 \text{ inches}$$

This is quite small, perhaps suitable for a pilot plant, but not for industrial use. Such small cyclones are regularly used in gas sampling.

9.10** (a) See example 9.5. The only change is that the inlet width is 1/2 of the one there so that the answer is that answer multiplied by $2 = 3.28$ μ.

(b) Using Equation 9.21 $$\eta = \frac{\left(\frac{1}{3.28}\right)^2}{1+\left(\frac{1}{3.28}\right)^2} = 0.085 = 8.5\%$$

9.11 Here the new inlet width is 50% of the old inlet width, the inlet area is 1/4 that of the old design, so that the new inlet velocity is 4 times the old inlet velocity .

$$D_{c_{new}} = D_{c_{old}}\sqrt{\frac{W_{i\,new}}{W_{i\,old}} \cdot \frac{V_{c_{old}}}{V_{c_{new}}}} = 10\mu\sqrt{\frac{1}{2}\cdot\frac{1}{4}} = 3.54\ \mu$$

The pressure drop, which is proportional to V^2 will increase by a factor of 16!

9.12 Starting with the equation in Example 8.4, substituting the centrifugal acceleration for that of gravity, we find

$$V_t = \left[\frac{\frac{4}{3}D\rho_p}{C_D\rho_f} \cdot \frac{V_c^2}{r}\right]^{\frac{1}{2}}$$

substituting this instead of Equation 9.15 into Equation 9.16, and simplifying, we find

$$\eta = \frac{N\pi}{W_i} \cdot \left[\frac{8D_0 D\rho_p}{3C_D\rho_f}\right]^{\frac{1}{2}}$$

It seems surprising that the efficiency does not depend on the gas velocity, but it makes sense. The radial outward velocity is proportional to the circumferential velocity, but so is the time spent in the cyclone, so these effects cancel (see Prob. 9.13). We also see that the drag on the particle is proportional to V^2 in this case compared to V^1 in the standard case.

This equation is not widely used, but it is probably the appropriate one to use for the large cyclones used to collect very large particles, e.g. sawdust in wood shops, wheat grains from pneumatic transport of wheat.

9.13 From Equation 9.15 $V_r = \dfrac{D^2\rho_p V_c^2}{18\mu r}$

The time spent in one turn is $\Delta t = \dfrac{2\pi r}{V_c}$ so that

$$\Delta r = V_r\Delta t = \frac{D^2\rho_p V_c}{\mu}\cdot\frac{\pi}{9} = \frac{(10\text{ E - 6m})^2 \cdot \left(2000\frac{\text{kg}}{\text{m}^3}\right)\cdot\left(20\frac{\text{m}}{\text{s}}\right)}{1.8\text{ E - 5}\frac{\text{kg}}{\text{m}\cdot\text{s}}}\cdot\frac{\pi}{9}$$

$$= 7.76\text{ E} - 4\text{ m} = 776\ \mu$$

This is not very far; it is hard to collect 1 μ particles in a reasonable size cyclone. This calculation ignores the increase in radius during one turn. A student worked that out, found that it made little difference.

9.14 See the solution to and comments on Problem 9.8 Using that solution

$$p_N = (p_1)^n; \quad N = \frac{\ln p_N}{\ln p_1} = \frac{\ln 0.005}{\ln 0.5} = 7.64 \approx 8$$

This is not a practical choice. Instead we should choose a smaller cyclone with a much lower cut diameter.

9.15** In Equation 9.20 we have $N = 0.25$, $W_i = 1$ cm, and

$$V = \frac{10^3 \frac{\text{cm}^3}{\text{s}}}{1\,\text{cm}^2} = 10^3 \frac{\text{cm}}{\text{s}} = 10 \frac{\text{m}}{\text{s}} \quad \text{so that}$$

$$D_c = \left[\frac{(9)\cdot(0.01\,\text{m})\cdot\left(1.8\,\text{E-}5 \frac{\text{kg}}{\text{m}\cdot\text{s}}\right)}{(2\pi)\cdot(0.25)\cdot\left(10\frac{\text{m}}{\text{s}}\right)\cdot\left(2000\frac{\text{kg}}{\text{m}^3}\right)} \right] = 7.18\,\text{E-}6\ \text{m} = 7.2\ \mu$$

This is only approximately correct, because the geometry of the nasal passage is much more complex than that of a cyclone, and the dimensions are only approximately correct. The observed value is $5\ \mu$. After the incoming air leaves the nasal passages, it has further removal of particles by collection in the trachea and bronchi. Only particles about 2-3 μ diameter or smaller get into the lungs.

9.16 The cut diameter is inversely proportional to the square root of the velocity, so the cut diameter will be $10\ \mu/\sqrt{2} = 7.07\ \mu$. As discussed later, pressure drop is proportional to the square of the velocity, so it will increase by a factor of 4.

9.17** From Equation 9.21 we find $\eta = \left(\frac{D}{D_c}\right) / \left(1 + \left(\frac{D}{D_c}\right)^2\right)$

so that

D, μ	D/D_c	η	p	Δw	Δwp	$\sum \Delta wp$
1	0.2	0.0385	0.962	0.33	0.317	0.317
5	1.0	0.5	0.5	0.33	0.165	0.482
10	2	0.8	0.2	0.34	0.068	0.550

The collection efficiency is 1-0.55 = 0.45 = 45%. The 1 μ particles almost all pass through, the 5 μ particles are 50% collected and the large particles are 80% collected.

9.18 First we assume a laminar boundary layer, for which we compute the thickness from Blasius solution at a downwind distance of $5 \cdot \pi \cdot D \cdot = 15.7$ ft,

$$\delta = 5\cdot\left(\frac{\nu x}{V_\infty}\right)^{\frac{1}{2}} = 5\cdot\left(\frac{1.613\,\text{E-}4\,\frac{\text{ft}^2}{\text{s}}\cdot 15.7\ \text{ft}}{60\,\frac{\text{ft}}{\text{s}}}\right)^{\frac{1}{2}} = 0.032\ \text{ft} = 0.39\,\text{in}$$

This is the value at the end of the five turns around the cyclone. Then we compute the corresponding Reynolds number, finding

$$\mathcal{R}_x = \frac{xV}{\nu} = \frac{(15.7\,\text{ft})\cdot\left(60\,\frac{\text{ft}}{\text{s}}\right)}{1.613\,\text{E}-4\,\frac{\text{ft}^2}{\text{s}}} = 5.8\times10^6$$

The laminar-turbulent transition typically occurs at values between 0.35 and $2.8\cdot10^6$, so we would expect that the transition would occur some where along the path, perhaps halfway along. The thickness of the corresponding turbulent boundary layer would be

$$\delta = 0.37x\left(\frac{\nu}{V_\infty x}\right)^{\frac{1}{5}} = 0.37\cdot\frac{\left(1.613\,\text{E-}4\,\frac{\text{ft}^2}{\text{s}}\right)}{60\,\frac{\text{ft}}{\text{s}}\cdot15.7\,\text{ft}} = 0.26\,\text{ft} = 0.186\text{ in}$$

Thus, in this simplified treatment, we would assume that near the inlet there would be a very thin laminar boundary layer, and near the bottom of the cyclone there would be a much thicker turbulent boundary layer.

This slow moving boundary layer probably improves the collection efficiency, by lessening the likelihood that the turbulence in the main flow will re-entrain the particles once they reach the wall.

9.19 The tangential velocity curves can be idealized as shown at the right. The section near the wall behaves as a free vortex, while that in the center behaves as a forced vortex. Both of these terms are described in any fluid mechanics text. These correspond to the fact that in the outer section there is a net inward flow, while in the inner section there is not.

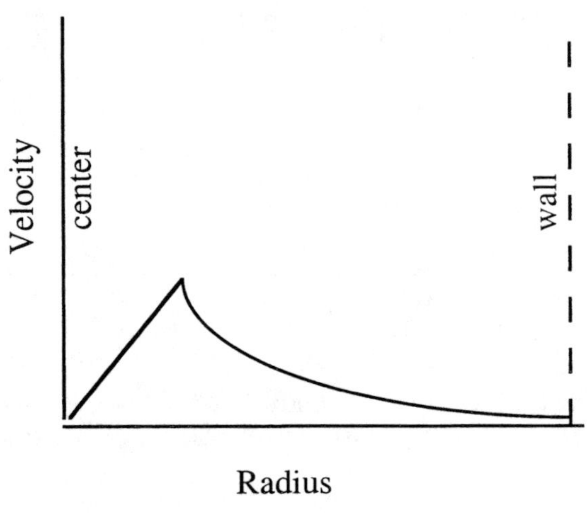

With MacDraw I cannot make the curve on the right appear as a segment of a rectangular hyperbola, which it is.

9.20 (a) We begin with a force balance on a particle, using the Stokes drag force given by Equation 8.3. Here the $V = V_g - V_p$, $\frac{dV_p}{dt} = \frac{18\mu(V_g - V_p)}{D^2\rho C}$; let $C = 1.00$

Then we separate variables and integrate from start to finish

$$\int_s^f \frac{dV_p}{V_g - V_p} = \frac{18\mu}{D^2\rho}\int_s^f dt;\ \ln\frac{(V_g - V_p)}{(V_g - V_p)_0} = -\frac{18\mu\Delta t}{D^2\rho}$$

The minus sign appears because of the - V_p in the denominator of the integral.

(b) Inserting values we have

$$\Delta t = \frac{(10\text{ E-6 m})^2 \cdot \left(2000 \frac{\text{kg}}{\text{m}^3}\right)}{(18)\cdot\left(1.8\text{ E-5} \frac{\text{kg}}{\text{m}\cdot\text{s}}\right)} \ln 100 = 2.84\text{ E-5 s}$$

One sees that the particle accelerates very rapidly (because it is so small and has so little mass) and practically moves with the fluid. Compare this calculation with the Stokes Stopping Distance calculation in Example 8.5. There we substituted for *dt* in terms of *V* and *dx* and found a distance instead of a time.

9.21 (a) $W_i = \frac{2\pi D^2 N V_c \rho_p}{9\mu}$

$$= \frac{2\pi\cdot(1\text{ E-6 m})^2 \cdot 5 \cdot 20 \frac{\text{m}}{\text{s}} \cdot \frac{2000\text{ kg}}{\text{m}^3}}{(9)\cdot\left(1.8\text{ E-5} \frac{\text{kg}}{\text{m}\cdot\text{s}}\right)} = 7.76\text{ E-3 m} = 7.76\text{ mm}$$

$D_o = 4\ W_i = 31\text{ mm} = 1.22\text{ inch}.$

(b) $Q = VW_iH = VW_i \cdot 2W_i = 2VW_i^2 = 2\cdot 20\frac{\text{m}}{\text{s}}\cdot(7.76\text{ E-3 m})^2 = 2.41\text{ E-3 }\frac{\text{m}^3}{\text{s}}$

Such devices are used for sampling purposes, but the volume of gas processed is small enough that they have no air pollution control application.

9.22

$$Q = VW_iH = 2VW_i^2; \quad W_i = \sqrt{\frac{Q}{2V}}; \quad W_i = \frac{2\pi D^2 N V_c \rho_p}{9\mu} = \sqrt{\frac{Q}{2V_c}}$$

squaring the rightmost equation and solving for *V*

$$V = \left(\frac{Q}{2 \cdot \left(\frac{2\pi D^2 N V_c \rho_p}{9\mu} \right)^2} \right)^{1/3}$$

$$= \left(\frac{1\,\mathrm{E}\text{-}5 \frac{\mathrm{m}^3}{\mathrm{s}}}{2 \cdot \left(\frac{(1\,\mathrm{E}\text{-}6\,\mathrm{m})^2 \cdot 2\pi \cdot 5 \cdot \left(\frac{2000\,\mathrm{kg}}{\mathrm{m}^3} \right)}{(9) \cdot \left(1.8\,\mathrm{E}\text{-}5 \frac{\mathrm{kg}}{\mathrm{m} \cdot \mathrm{s}} \right)} \right)^2} \right)^{1/3} = 33.2 \frac{\mathrm{m}}{\mathrm{s}}$$

From Eq. 9.22

$$\Delta P = \frac{K \rho_g V_c^2}{2} \approx \frac{8 \cdot 1.2 \frac{\mathrm{kg}}{\mathrm{m}^3} \cdot \left(33.2 \frac{\mathrm{m}}{\mathrm{s}} \right)^2}{2} = 5.3\ \mathrm{kPa} = 0.77\ \mathrm{psig}$$

9.23** We write the total annual cost as

$$TAC = \text{Power Cost} + \text{Capital Cost} = C_1 Q \Delta P + C_2 C_3 N D_0^{\ 2}$$

Here N is the number of cyclones in parallel, not the number of turns in an individual cyclone. From Equation 9.22 we know that $\Delta P = \frac{K V_{\text{in}}^{\ 2} \rho_g}{2}$. And we know that the number of cyclones in parallel is $N = \frac{Q}{A_{\text{inlet}} V_{\text{inlet}}}$ Substituting these in the total annual cost equation and simplifying,

$$TAC = \frac{C_1 Q K V_{\text{in}}^{\ 2} \rho_g}{2} + \frac{C_2 C_3 D_0^{\ 2} Q}{A_{\text{in}} V_{\text{in}}}$$

Then we take the derivative of TAC with respect to V_{in}, set it equal to zero, and

$$\frac{d(TAC)}{dV_{\text{in}}} = 0 = \frac{2}{2} K C_1 \rho_g V_{\text{in}} - \frac{C_2 C_3 D_0 Q}{A_{\text{in}} V_{\text{in}}^2}; \quad V_{\text{in}} = \left[\frac{C_2 C_3}{K C_1 \rho_g} \cdot \frac{D_0^{\ 2}}{A_{\text{in}}} \right]^{\frac{1}{3}}$$

From Figure 9.4 $\frac{D_0^{\ 2}}{A_{\text{in}}} = 8$, so that

$$V_{in} = \left[\frac{8}{K}\cdot\frac{C_2C_3}{C_1\rho_g}\right]^{\frac{1}{3}}; \text{ for } K = 8, \quad V_{in} = \left[\frac{C_2C_3}{C_1\rho_g}\right]^{\frac{1}{3}}$$

This shows the proper relationship between economic velocity and the cost factors. If we increase the power cost without changing the other costs, the C_1 increases, and the economic velocity decreases. If we increase C_2 or C_3 that raises the capital cost, and the economic velocity increases. A design with a higher K will have a lower economic velocity.

Once the economic velocity is known, from that information and the required cut diameter and gas properties, one may compute the required value of D_o. Then one may compute the number of cyclones in parallel required to handle the design Q.

If one puts reasonable values of the parameters into this equation one finds that the economic velocity is close to 60 ft/s, for almost any size cyclone.

9.24 By straightforward substitution, we find that for a 1 μ particle, $\Psi = 0.00058$ and $\eta =$ 0.494. This collection efficiency is substantially higher than is calculated in Example 9.4; it suggests a cut diameter for a conventional cyclone of about 1 μ. Ref [1] shows computations by Lieth and Licht's method, showing that at higher collection efficiencies the values are similar to those computed in this chapter. Only limited comparisons between experiments and calculations are available to settle the question of which approach is more satisfactory.

9.25 (a) $p = 1 - \frac{(D/D_c)^2}{1-(D/D_c)^2}$; Put this over a common denominator, cancel like terms and expand by binomial theorem to find the result shown.

(b) By simple series expansion we find the result shown.

(c) In example 9.6 the penetration is 0.54 for particles with $D_{cut}/D_{mean} = 0.275$, so that a penetration of 0.5 corresponds approximately to $D_{cut}/D_{mean} = 0.25$. The σ is stated as 1.25, so that exp $\sigma = 3.49$. Then simple chart reading on Fig. 9.30 gives an overall penetration of 0.14

(d) Example 9.6 was all in terms of physical diameter, not aerodynamic diameter. As long as the particle size distribution and the collection efficiency relation both in the same dimension (either physical or aerodynamic diameters) both the procedure in Example 9.6 and Figure 9.30 work properly.

(e) If $p = \exp(-AD^2)$ and 0.5 = $\exp(-A\cdot(5\mu)^2)$ then $A = 0.02773/\mu^2$. If one repeats the spreadsheet calculation leading to Tab. 9.1, using this relation for p instead of the relations shown there, one does indeed find an overall penetration of 0.14, the same as one finds from Fig. 9.30 (which was certainly made up the same way).

9.26 Changing from gravity settler to ESP notation, we have

$$V_t = w; \quad V_{avg} = \frac{Q}{Hh}; \quad h = \frac{A}{L}; \quad V_{avg} = \frac{QL}{AH}$$

$$\eta = \frac{LV_t}{HV_{avg}} = \frac{LwHA}{HQL} = \frac{wA}{Q}$$

so that H is implicitly present in Q/A. In addition H appears implicitly because E V/H which influences w.

9.27** Before the change $p = 0.05 = \exp\left(\frac{-wA}{Q}\right); \quad \frac{wA}{Q} = 3.0.$

After the change

$$\frac{wA}{Q} = \frac{3}{2} = 1.5; \quad p = \exp(-1.5) = 0.22; \quad \eta = 1 - p = 0.78$$

This surprises students and leads to more discussion that one might expect for such a simple plug-in problem. I ask what would happen if we had the block flow assumptions, answer the efficiency would drop by a factor of 2. Then I ask them what would happen for the block flow assumptions for $(wA/Q) = 3$? That leads to a nonsense answer; the block flow assumption is practically never used for ESPs.

9.28** (a) $p = 0.05$ means $\frac{wA}{Q} = 3.0,$ For $p = 0.01, \frac{wA}{Q} = 4.6$

So the new area is

$$\frac{A_{\text{new}}}{A_{\text{old}}} = \frac{\left(\frac{wA}{Q}\right)_{\text{new}}}{\left(\frac{wA}{Q}\right)_{\text{old}}} = \frac{4.6}{3.0} = 1.54$$

We need 5.4 new units, or using integral values, 6.

(b) It makes no difference. If one puts the six new ones in parallel with the 10 existing ones, then each of the 16 will have 10/16 of the original flow rate, and the required efficiency will be met. If the 6 new ones are placed in parallel with each other, but in series with the existing 10, either upstream or downstream, the required efficiency will be met. It is a worthwhile exercise to ask the students to sketch the situation, to see that. In a real situation the decision would most likely be made based on the difficulty and cost of the ductwork, foundations and controls for the two arrangements. Most often the series arrangement would be cheaper, but not always.

9.29 $p = 0.05$ means $\left(\frac{wA}{Q}\right) = 3,$ so that the new $\left(\frac{wA}{Q}\right) = 2 \times 3 = 6$

$$p = \exp(-6) = 0.0025, \ \eta = 1 - 0.0025 = 99.75\%$$

9.30** From Table 9.2 we read that for multiple hearth roaster, a typical industrial value of the drift velocity is 0.26 ft/s. Then

$$p = 0.005 = \exp\left(\frac{-wA}{Q}\right); \quad \frac{wA}{Q} = -\ln(0.005) = 5.30$$

$$A = \frac{wA}{Q} \cdot \frac{Q}{w} = 5.30 \cdot \frac{1000 \frac{\text{ft}^3}{\text{min}}}{0.26 \frac{\text{ft}}{\text{s}}} \cdot \frac{\text{min}}{60\,\text{s}} = 3.4\text{ E3 ft}^2$$

9.31 $p = 0.05, \quad \dfrac{wA}{Q} = 3.0$; for $p = 0.01, \quad \dfrac{wA}{Q} = 4.605$

$$\left(\frac{E_{new}}{E_{old}}\right)^2 = \frac{w_{new}}{w_{old}} = \frac{4.605}{3.00}; \quad \frac{E_{new}}{E_{old}} = \sqrt{\frac{4.605}{3.00}} = 1.24$$

We must increase the voltage by 24%. If the current voltage is 40 kV this means we must increase it to 50 kV. The solid state electronic controls are better than the older electromechanical ones, so we can use higher voltages. I don't know if we can make this large an increase.

9.32 $p = 0.05$ means $\dfrac{wA}{Q} = 3.0, \qquad \dfrac{Q_2}{Q_1} = \dfrac{T_2}{T_1} = \dfrac{250 + 460\ °R}{600 + 460\ °R} = 0.6698$

For the cooled gas

$$\frac{wA}{Q} = \frac{3.0}{0.6698} = 4.473; \quad p = \exp(-4.473) = 0.0114; \quad \eta = 99.86\%$$

This is a satisfactory problem to let them see the effect of cooling on gas flow rate, and thus residence time of the gas in the ESP. However the assumption that the drift velocity does not change is only fair. There are well-known examples in which it changes significantly with temperature changes of this magnitude.

9.33 The volume between 2 plates, (See Fig 9.7) is $V = 2HLh$. The time the gas spends passing through it is

$$\Delta t = \frac{V}{Q} = \frac{2Lh \cdot H}{Q} = \frac{A}{Q}H; \quad \text{so that } \frac{A}{Q} = H\Delta t$$

The problem does not call for it, but as an example, a modern ESP may have

$$\frac{A}{Q} \approx \frac{400\text{ ft}^2}{1000\text{ cfm}} = 0.4\frac{\text{min}}{\text{ft}}$$

If the plate spacing is $2H = 1$ ft; $H = 0.5$ ft; $\quad \Delta t = 0.4\dfrac{\text{ft}}{\text{min}} \cdot 0.5\text{ ft} = 0.2\text{ min} = 12\text{ s}$

9.34** $\dfrac{w_{\text{new}}}{w_{\text{old}}} = \left(\dfrac{E_{\text{new}}}{E_{\text{old}}}\right)^2 = \left(\dfrac{50\text{ kV}}{40\text{ kV}}\right)^2 = 1.25^2 = 1.56$

$$p_{\text{old}} = 0.05; \quad \left(\frac{wA}{Q}\right)_{\text{old}} = 3.00; \quad \left(\frac{wA}{Q}\right)_{\text{new}} = 3.00 \cdot 1.56 = 4.68$$

$$p_{\text{new}} = \exp(-4.68) = 9.27\text{ E}-3; \quad \eta_{\text{new}} = 99.07\%$$

See the discussion in Prob. 9.31. I don't know if this is an accurate description of their ESP upgrade; I do know that they did such an upgrade in 1976.

9.35 $p = 0.05 = \dfrac{1}{3}(p_1 + p_2 + p_3) = \dfrac{1}{3}\left(\exp\left(\dfrac{-w_1A}{Q}\right) + \exp\left(\dfrac{-w_2A}{Q}\right) + \left(\dfrac{-w_3A}{Q}\right)\right)$

$$\text{let } \frac{w_3A}{Q} = \alpha; \quad \text{then} \quad 3\cdot(0.05) = \left[\exp\left(\frac{-10}{3}\alpha\right) + \exp\left(\frac{-7}{3}\alpha\right) + \exp(-\alpha)\right]$$

Using the solve routine on my hand calculator, I find that $\alpha = 1.97568$. Substituting back into the above expressions I find that the individual penetrations sum to 0.1500, and $p =$ 0.05. The individual efficiencies are 99.86, 99.00 and 86.13 %

This problem illustrates (in simplified form) that ESPs selectively capture the large particles, and that the exhaust gas has a particle size distribution far different from that of the inlet gas.

9.36 Starting with Eq. 7.10, and substituting Eq. 9.30 for the penetration, we find

$$p_{\text{overall}} = \int p(D)\frac{d\phi}{dD}dD = \int \exp\left(\frac{-C_2DA}{Q}\right)C_1 D dD; \quad \text{let } \frac{-C_2A}{Q} = \alpha$$

$$p_{\text{overall}} = C_1\int_0^{D_{\text{mean}}} D\exp(\alpha D)dD = C_1\left|\frac{(\alpha D-1)}{\alpha^2}\exp(\alpha D)\right|_0^{D_{\text{mean}}}$$

$$= \frac{C_1}{\alpha^2}\left[\alpha D_{\text{mean}}\exp(\alpha D_{\text{mean}}) - \exp(\alpha D_{mean}) + 1\right]$$

The resulting equation has little intuitive content. However if one chooses $p = 0.1$ at $A = A_0$, and computes the relation between p and A, one finds that this relation gives practically the same values as Equation 9.31, with $k = 0.5$. Thus, although this equation has some intellectual basis, it is harder to use, and give the same result as the simple data fitting Equation 9.31.

9.37(a) For the existing precipitator, $p = 0.1$

$$w = 0.506 + 0.0356\ln(0.1) = 0.424; \quad \frac{A}{Q} = -\frac{\ln p}{w} = -\frac{\ln 0.1}{0.424} = 5.43\frac{\text{s}}{\text{ft}}$$

For the new one, $p = 0.01$

$$w = 0.506 + 0.0356\ln(0.01) = 0.342; \quad \frac{A}{Q} = -\frac{\ln(0.01)}{0.342} = 13.46\frac{\text{s}}{\text{ft}}$$

This is 2.5 times the area for the 90 % efficient precipitator, which is not as conservative as Equation 9.31 with $k = 0.5$, which requires 4 times as much area.

(b) This requires trial and error on a spreadsheet. Calculate *A/Q* for the simple DA equation, using the changing values of *w* in this problem, and then compute their ratio to the values one would calculate using the *w* corresponding to 95% efficiency. Repeat the computation using Equation 9.31, with *k* taken as a variable. By trial and error find the value of *k* which makes the ratios the same. I find that there is good agreement for *k*= 0.73, which is, as above, less conservative than $k = 0.50$.

9.38 (a) This is Example 9.15 extended to three chambers. For $p_o = 0.05$, $\dfrac{wA}{Q} = 3.00$

$$p_{new} = \sum \frac{Q_i}{Q_T}\exp\left(-3.00\frac{\left(\frac{Q_T}{3}\right)}{Q_i}\right) = 0.5\exp\left(\frac{-1}{0.5}\right) + 0.3\exp\left(\frac{-1}{0.3}\right) + 0.2\exp\left(\frac{-1}{0.2}\right) = 0.920$$

(b) By the same logic

$$\eta_{\text{overall}} = \sum \frac{Q_i}{Q_T}\eta_i; \quad \eta = 0.70\frac{Q_i}{Q_{\text{original}}} = 0.70\frac{Q_i}{Q_T/2}$$

$$\eta_{\text{overall}} = \frac{Q_T^2}{Q_t^2}\left[\frac{0.7}{0.5}\left(\frac{2}{3}\right)^2 + \frac{0.7}{0.5}\left(\frac{1}{3}\right)^2\right] = 0.777 = 77.7\%$$

(c) Here we see that for ESPs maldistribution lower efficiency, while for cyclones (or multiclones, which correspond to a multi-section ESP) maldistribution increases efficiency (while raising pressure drop or lowing flowrate). The reason is that increasing the velocity through a section of an ESP lowers its efficiency, while increasing the velocity through a cyclone increases its efficiency. In an ESP with maldistribution, more of the gas flows through the low efficiency section. In a set of cyclones in parallel with maldistribution, more of the gas flow goes through the high efficiency section.

9.39 Let $wA = \beta$, $\quad \eta_0 = 1 - \exp\left(\dfrac{-\beta}{(Q/2)}\right)$

$$\eta_{new} = \frac{Q_1}{Q}\left(1-\exp\left(\frac{-\beta}{Q_1}\right)\right) + \frac{Q_2}{Q}\left(1-\exp\left(\frac{-\beta}{Q_2}\right)\right)$$

$$\frac{\eta_{new}}{\eta_0} = \frac{\frac{Q_1}{Q}\left(1-\exp\left(\frac{-\beta}{Q_1}\right)\right) + \left(1-\frac{Q_1}{Q}\right)\cdot\left(1-\exp\left(\frac{-\beta}{Q-Q_1}\right)\right)}{1-\exp\left(\frac{-\beta}{Q/2}\right)}$$

The plot is sketched at the right. It is symmetric about $Q_1/Q = 0.5$, and the curve is higher for higher values of wA. (ESPs with high collection efficiency suffer less loss of efficiency with flow maldistribution than those with lower collection efficiencies).

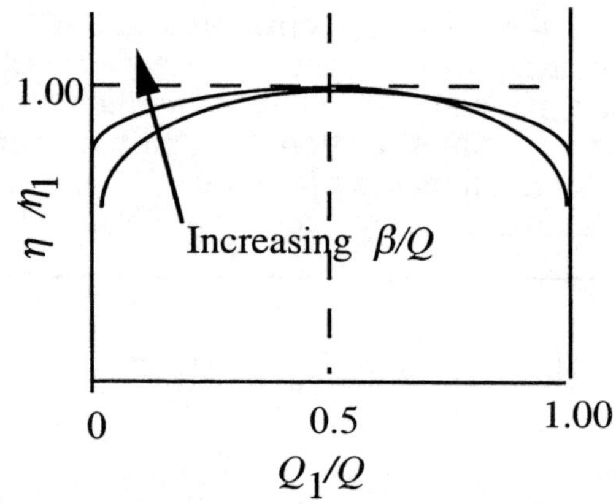

9.40 From Example 7.6, the SO_2 content is 1130 ppm. Thus 10 to 30 ppm of SO_3 correspond to 0.8 to 2.6% of the SO_2 in the gas stream. EPRI is currently (1999) running a pilot plant in which they circulate a part of the exhaust gas from a power plant over an oxidizing catalyst to convert the contained SO_2 to SO_3 and then mix that with the gas going to the ESP. That seems likely to be cheaper than using purchased SO_3.

9.41 (a) $$p = \exp\left(-\frac{wA}{Q}\right); \quad w = (-\ln p)\cdot\frac{Q}{A}$$

$$w = (-\ln 0.05)\cdot\frac{\text{ft}}{0.035\text{ min}}\cdot\frac{\text{min}}{60\text{ s}} = 1.42\frac{\text{ft}}{\text{s}}$$

(b) The experimental drift velocity is strongly influenced by re-entrainment of the captured particles. In a wet ESP there is practically no re-entrainment, so the observed drift velocities are 3 times as large as in dry ESPs.

(c) Dry ESPs are simpler and cheaper to buy and operate, because of the extra cost and complexity of the water injection, removal and treatment systems in a wet ESP. For that reason there are more dry ESPs than wet ESPs in industry.

9.42

(a)

$$\Delta P = V_s \mu \left[\frac{\Delta z}{k} + \alpha \right]$$

$$\frac{d(\Delta P)}{dt} = \frac{V_s \mu}{k} \frac{d(\Delta x)}{dt}$$

$$\Delta x = \frac{m / A}{\rho_c}; \quad \frac{d(\Delta x)}{dt} = V_s W \eta$$

$$\frac{d\Delta P}{dt} = \frac{V_s^2 \mu W \eta}{k} = \text{constant}$$

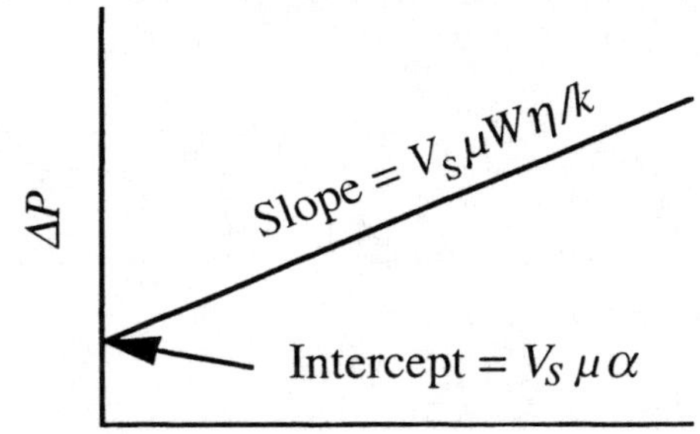

(b) Starting with Eq. 9.40 and

$$\Delta x = \frac{V}{A} \cdot W\eta,$$

$$\Delta x^2 \left(\frac{\mu W}{2k} \right) \cdot \left(\frac{1}{W\eta} \right)^2 + \Delta x \left(\frac{\mu \alpha}{W\eta} \right) = t \Delta P$$

See Ref [14].

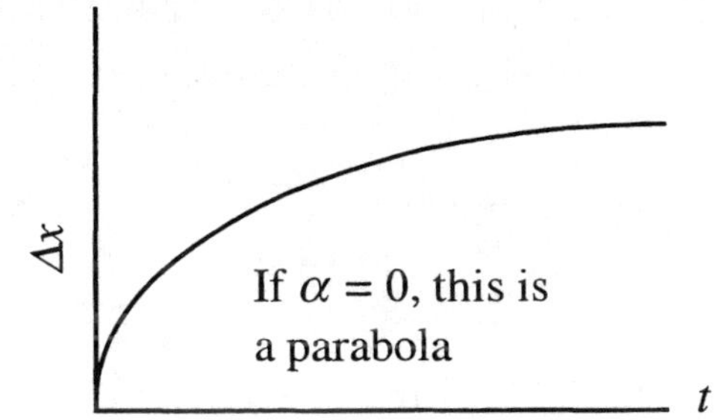

(c) By inspection

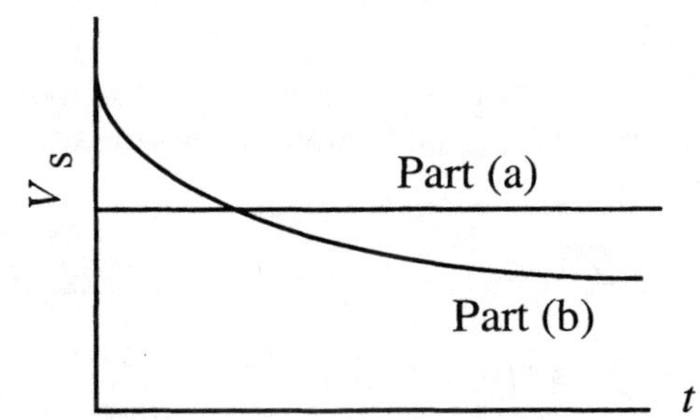

The point of this problem is to make the students look at the expected behavior of a surface filter, for the two most common kinds of feed flows. If there are civil engineers in the class, then it is appropriate to also sketch the behavior for a compressible filter cake, which is found regularly in water treatment plants. The appropriate plot is one of V_S vs P at constant cake thickness. For an incompressible cake it shows a straight line through the origin; for a compressible cake the curve has a maximum, going to zero at high values of P.

9.43 (a) By inspection $K_f + W_r K_2 = 0.1288 \frac{\text{in H}_2\text{O}}{\text{ft / min}}$

(b) By inspection $K_2 = 140.59 \left(\frac{\text{in H}_2\text{O}}{\text{ft / min}} \cdot \frac{\text{ft}^2}{\text{lbm}} \right)$

(c) $\Delta W = c V_s \Delta t = 13 \frac{\text{gr}}{\text{ft}^3} \cdot 11.3 \frac{\text{ft}}{\text{min}} \cdot \frac{60 \text{ min}}{3.3} = 24.65 \frac{\text{gr}}{\text{ft}^2} = 0.00352 \frac{\text{lbm}}{\text{ft}^2}$

(d) Here we assume that at the start of a cycle $W = 0$, so that at the end of a cycle W = the value calculated in part (c). Then

$$\Delta P = 11.3 \frac{\text{ft}}{\text{min}} \cdot (140.59 \cdot 0.00352 + 0.1288) \frac{\text{in H}_2\text{O}}{\text{ft / min}} = 7.04 \text{ in H}_2\text{O}$$

(e) $\Delta P = 11.3 \frac{\text{ft}}{\text{min}} \cdot (0.1288) \frac{\text{in H}_2\text{O}}{\text{ft / min}} = 1.455 \text{ in H}_2\text{O}$

$$W_r = \frac{\Delta P / V}{K_2} = \frac{1.455 \text{ in H}_2\text{O} / 11.3 \text{ (ft / min)}}{\left(140.59 \frac{\text{in H}_2\text{O} / (\text{ft / min})}{\text{lbm / ft}^2}\right)} = 9.15 \text{ E - 4} \frac{\text{lbm}}{\text{ft}^2}$$

(f)

$$WK_2 = \mu \frac{\Delta x}{k}; \quad W = \rho \Delta x: \;\; k = \frac{\mu}{\rho K_2}$$

At 340°F = 171°C, the air viscosity correlation inside the back cover shows μ 0.025 cp. The bed density is $\rho = 2.00 \cdot 0.5 \cdot 62.3 \quad 62.3$ lbm/ft³. Thus

$$k = \frac{0.025 \text{ cp}}{\left(62.3 \frac{\text{lbm}}{\text{ft}^3}\right) \cdot \left(140.59 \left(\frac{\text{in H}_2\text{O}}{\text{ft / min}} \cdot \frac{\text{ft}^2}{\text{lbm}}\right)\right)} \cdot \frac{\text{in H}_2\text{O}}{0.0361 \text{ lbf / in}^2}$$

$$\cdot \frac{2.09 \text{ E - 5 lbf} \cdot \text{s}}{\text{ft}^2 \cdot \text{cp}} \cdot \frac{\text{min}}{60 \text{ s}} \cdot \frac{\text{ft}^2}{144 \text{ in}^2} = 1.78 \text{ E - 13 ft}^2 = 0.0167 \text{ darcy}$$

The darcy is the unit of permeability in petroleum reservoir engineering and in some ground water flow calculations. This corresponds to the permeability of a more-or-less averages sandstone.

(g) Eq. 12.14 of Ref. [13] , written for horizontal laminar flow is

$$-\Delta P = \frac{150\mu \cdot (1-\varepsilon)^2 \Delta x V_s}{D_p^3 \varepsilon^3} \quad \text{and Eq. 9.72 can be written } k = \frac{V_s \Delta x \mu}{-\Delta P}$$

If we solve the first of these for V_S and substitute it in the second we find

$$k = \frac{D_p^3 \varepsilon^3}{150\,(1-\varepsilon)^2} = \frac{(4\,\text{E} - 6\text{ m})^2 \cdot 0.5^3}{150 \cdot 0.5^2} = 5.33 \text{ E - 14 m}^2 = 5.73 \text{ E - 13 ft}^2$$

which is within a factor of 3 of the experimental value. Given the uncertainties in particle diameter and porosity, this is better agreement than we should expect, and probably a coincidence.

9.44 (a) Here $W = 0$, so that $V_s = \frac{\Delta P}{K_r} = \frac{7 \text{ in H}_2\text{O}}{0.1288 \text{ in H}_2\text{O} / (\text{ft / min})} = 54.3 \text{ ft / min}$

(b) We assume $\eta = 1.00$, $\frac{dW}{dt} = cV_s$; $V_s = \frac{\Delta P}{AW + B}$ where A and B are constants, then

$$\frac{dW}{dt} = \frac{c\Delta P}{AW + B}; \quad (AW + B)dW = c\Delta P dt$$

$$\frac{AW^2}{2} + BW = c\Delta P\Delta t; \quad W^2 + \frac{2B}{A}W - \frac{2c\Delta P\Delta t}{A} = 0$$

This is a quadratic equation which we can solve for W.

$$\frac{2B}{A} = \frac{2 \cdot 0.1288}{140.59} = 1.83 \text{ E - 3}; \quad \frac{2c\Delta P\Delta t}{A} = \frac{2 \cdot 7 \cdot \left(\frac{60}{3.3}\right)}{140.59} = 3.36 \text{ E - 5},$$

all in consistent units,

$$W^2 + 1.83 \text{ E - 3}W - 3.66 \text{ E - 5} = 0; \quad W = 4.88 \text{ E - 3}\frac{\text{lbm}}{\text{ft}^2}$$

which is slightly higher than the answer to Prob. 9.43 (c), reflecting the fact that the flow rate-time plot is parabolic, not linear.

(c) $V_s = \dfrac{7}{(140.59)\cdot(4.88 \text{ E - 3}) + 0.1288} = 8.58\dfrac{\text{ft}}{\text{s}}$ which is less than the average velocity, as we would expect.

9.45**(a)

$$A = \frac{Q}{Q/A} = \frac{1.5 \text{ E6}\dfrac{\text{ft}^3}{\text{min}}}{5\dfrac{\text{ft}}{\text{min}}} = 3.00 \text{ E5} \,\text{ft}^2$$

(b) Each bag has a $a = \pi DL = \pi \cdot 1\,\text{ft} \cdot 40\,\text{ft} = 125.66\,\text{ft}^2$

so $N = \dfrac{A}{a} = \dfrac{3.00 \text{ E5}\,\text{ft}^2}{125.7\,\text{ft}^2} = 2388$ bags

This is a simple "order of magnitude" problem, showing that while the filter size required is large, it is not impossibly large.

9.46 Starting with Eq. 9.39 and $\alpha = 0$,

$$\Delta P = \frac{\mu\Delta x}{k}\cdot\frac{Q}{A} = \Delta P_0 - bQ$$

$$Q\left(b + \frac{\mu\Delta x}{kA}\right) = \Delta P_0 = \frac{A}{W}\left(b + \frac{\mu\Delta x}{kA}\right)\frac{d\Delta x}{dt}; \quad b\Delta x + \frac{\mu}{kA}\frac{(\Delta x)^2}{2} = \frac{W\Delta P_0}{A}\Delta t$$

$$\Delta x = \frac{m}{A\rho_{cake}}; \quad m^2\left(\frac{\mu}{2kA}\cdot\frac{1}{A^2\rho_{cake}^2}\right) + m\cdot\frac{b}{A\rho_{cake}} = \frac{W\Delta P_0}{A}\Delta t$$

One may check by setting $b = 0$, and showing, after suitable algebra that this is the same as Equation 9.40. This equation shows that if $b > 0$, then the mass accumulated per unit time will be less than if $b = 0$. That is the real case for most industrial blowers.

9.47 (a) For all, $c_O = 0.8$ g/m^3. Then

V_s, m/min	c_o, g/m^3	p	η
3.35	0.16	0.2	80%
1.52	0.019	0.02375	97.6%
0.61	0.0009	1.125 E-3	99.89%
0.39	0.0006	7.5 E-4	99.925%

(b) $$\frac{\text{mass}}{\text{area}} = \rho_{particle} \cdot (1-\varepsilon)\Delta x = 140\frac{\text{g}}{\text{m}^2}$$

$$\Delta x = \frac{140\dfrac{\text{g}}{\text{m}^2}}{2\text{ E6}\dfrac{\text{g}}{\text{m}^3}\cdot(1-0.3)} = 1\text{ E - 4 m} = 0.1\text{ mm}$$

(c) $$\frac{\text{mass}}{\text{area}} = 140\frac{\text{g}}{\text{m}^2} = c\eta V_s \Delta t$$

assuming that η 1.00 for the whole process, we have

$$\Delta t = \frac{140\dfrac{\text{g}}{\text{m}^2}}{cV_s} = \frac{140\dfrac{\text{g}}{\text{m}^2}}{0.8\dfrac{\text{g}}{\text{m}^3}\cdot 0.39\dfrac{\text{m}}{\text{min}}} = 449\text{ min}$$

9.48 (a) Total Annual Cost = TAC = Power Cost + Capital Cost

$$TAC = C_1\Delta PQ + C_3A = C_1C_2\frac{Q}{A}\cdot Q + C_3A = \frac{C_1C_2Q^2}{A} + C_3A$$

$$\frac{d(TAC)}{dA} = \frac{-C_1C_2Q^2}{A^2} + C_3; \quad \text{set} = 0\text{, find} \left(\frac{A}{Q}\right)_{\text{optimum}} = \sqrt{\frac{C_1C_2}{C_3}}$$

(b) In that problem *Q/A* was equal to the cube root of a set of constants. Here it is proportional to the square root of a set of constants. The reason for the difference is that in Problem 9.23 the gas flow was turbulent, and hence the pressure drop was proportional to the square of the velocity, while here the flow (inside the filter cake) is laminar, and the pressure drop is proportional to the gas velocity.

(c) $$\left(\frac{Q}{A}\right)_{\text{opt}} = \sqrt{\frac{C_3}{C_1C_2}} = \left[\frac{\$2}{\text{ft}^2\text{ yr}}\cdot\frac{\text{ft / min}}{0.025\text{psi}}\cdot\frac{\text{kw yr}}{\$300}\cdot\frac{1000\text{w}}{\text{kw}}\cdot\frac{\text{ft lbf}}{1.355\text{ Ws}}\cdot\frac{\text{min}}{60\text{ s}}\cdot\frac{\text{ft}^2}{144\text{ in}^2}\right]^{\frac{1}{2}}$$

$$= 0.151\frac{\text{ft}}{\text{s}} = 9.06\frac{\text{ft}}{\text{min}} = 0.05\frac{\text{m}}{\text{s}}$$

This is close to the velocity used in these filters.

9.49

$$TAC = \frac{C_1 C_2 Q^2}{A} + C_3 A^{0.6}$$

$$\frac{d(TAC)}{dA} = \frac{-QC_1C_2}{A^2} + \frac{0.6C_3}{A^{0.4}}; \quad \text{set} = 0, \quad A = \left(\frac{Q^2 C_1 C_2}{0.6C_3}\right)^{\frac{1}{1.6}}$$

In this case we cannot write *Q/A* as a function of the constants alone, because of the non-linear cost relation ship. This is probably closer to actual costs than the linear relationship, so that as the size of the flow to be treated increases, the optimum gas velocity becomes smaller.

9.50 (a) $F = PA = 30 \text{ in } H_2O \cdot 100 \text{ ft}^2 \cdot \frac{144 \text{in}^2}{\text{ft}^2} \cdot \frac{62.4 \quad \text{psi}}{1728 \text{ in} H_2O} = 15{,}600 \text{ lbf}$

(b) For a rectangular wall, it makes little difference. For a cylinder it makes a big difference. Most baghouses have rectangular walls, and have steel frames strong enough to resist the force calculated in part (a).

(c) Many baghouses have safety vents which pop open if the pressure inside becomes significantly greater than the design pressure (or vacuum).

9.51**

$$N_s = \frac{\left(2000\frac{\text{kg}}{\text{m}^3}\right) \cdot (1\text{E-5 m})^2 \left(100\frac{\text{ft}}{\text{s}}\right)}{18 \cdot \left(1.8\text{E-5}\frac{\text{kg}}{\text{m} \cdot \text{s}}\right)} \cdot \frac{\text{m}}{3.28 \text{ft}} = 1.88$$

From Figure 9.18 we read. $n_T = 0.69$

In the class discussion this is good place to make sure they understand the swept area concept. This is also a good place to say a few words about Irving Langmuir; he was the first industrial Nobel Prize winner, (General Electric) who did many pioneering things in physical chemistry. It is also a good place to discuss icing of aircraft. That is only a problem at low altitudes. One regularly sees modern passenger planes being sprayed with hot de-icing solutions before takeoff in cold, wet weather. They get up through the icing region fast enough that they normally have little problem with it. The aircraft in World War II, when Langmuir and Blodget produced this famous plot, did not fly that high, and icing was a serious problem. (It was also a problem in carburetors, which the piston engines of that period had.) It is one of the most common causes of private air plane crashes.

9.52 This problem is identical to Example 9.18, with the particle diameter changed from 1 to 10 μ. That increases the separation number by a factor of 100 to 61.7, for which the target efficiency is 1.00. The collection rate will be 1 E-8 $g/m^2 \cdot s$.

9.53**

$$\eta_i = (1-0.8)\cdot(0.25) = 0.05; \quad p_i = 1 - n_i = 0.95$$

$$p_{\text{overall}} = p_i^{\,n} = (0.95)^{10} = 0.598$$

$$\eta_{\text{overall}} = 1 - p_{\text{overall}} = 0.4012 = 40\%$$

This is the logical place to discuss the distinction between depth filters and surface filters. This particular problem was chosen to correspond to a type of filter most of the students have actually seen. At this point I regularly bring to class a throwaway dust mask, which I pass around for the students to examine. It is also a depth filter.

9.54 Following example 9.18, we find the impaction target efficiencies, and then the collection efficiencies for each particle size as

$$\text{For } 10\mu,\ N_s = \frac{\left(2000\dfrac{\text{kg}}{\text{m}^3}\right)\cdot(10\ \text{E}-6\ \text{m})^2\cdot\left(1\dfrac{\text{m}}{\text{s}}\right)}{(18)\cdot\left(1.8\ \text{E-}5\dfrac{\text{kg}}{\text{m}\cdot\text{s}}\right)\cdot(100\ \text{E}-6\ \text{m})} = 6.172; \eta_t = 0.85,\quad \eta = 0.425$$

$$\text{For } 0.01\mu;\quad N_s = 6.172\cdot\left(\frac{0.01}{10}\right)^2 = 6\ \text{E-}6;\quad \eta_t = 0,\quad \eta = 0$$

For diffusional collection we follow Example 9.21 as follows

$$\text{For } 10\mu,\ C\approx 1;\quad \mathcal{D} = 2.8\ \text{E-}11\frac{\text{m}^2}{\text{s}}\cdot\frac{1}{1.16}\cdot\frac{1\,\mu}{10\,\mu} = 2.4\ \text{E-}12\frac{\text{m}^2}{\text{s}}$$

$$\text{For } 0.01\mu,\ C\approx 13.1;\quad \mathcal{D} = 2.8\ \text{E-}11\frac{\text{m}^2}{\text{s}}\cdot\frac{13.1}{1.16}\cdot\frac{1\,\mu}{0.01\,\mu} = 3.16\ \text{E-}8\frac{\text{m}^2}{\text{s}}$$

$$\text{For } 10\mu,\ \eta_{t\ \text{Diffusion}} = 0.0086\cdot\left(\frac{2.4\ \text{E-}12}{6\ \text{E-}10}\right)^{\frac{2}{3}}\cdot\left(\frac{1\ \text{E-}5}{1\ \text{E-}4}\right)^{\frac{1}{7}} = 6.8\ \text{E-}5\approx 0$$

$$\text{For } 0.01\mu,\ \eta_{t\ \text{Diffusion}} = 0.0086\cdot\left(\frac{3.16\ \text{E-}8}{6\ \text{E-}10}\right)^{\frac{2}{3}}\left(\frac{1\ \text{E-}5}{1\ \text{E-}7}\right)^{\frac{1}{7}} = 0.038$$

This illustrates that impaction works well for large particles, but not for small ones, and diffusion just the reverse.

9.55** (a) To get 40% overall efficiency, we need a target efficiency of 80%, which corresponds to a separation number of 3.4. That corresponds to

$$D_p = \left[\frac{N_s \cdot 18\mu D_b}{V\rho_p}\right]^{\frac{1}{2}} = \left[\frac{3.4\cdot 18\cdot\left(1.8\,\text{E-}5\dfrac{\text{kg}}{\text{m}\cdot\text{s}}\right)\cdot(1\,\text{E-}4\text{ m})}{\left(1\dfrac{\text{m}}{\text{s}}\right)\left(2000\dfrac{\text{kg}}{\text{m}^3}\right)}\right]^{\frac{1}{2}} = 7.4\,\text{E-}6\text{ m} = 7.4\ \mu$$

(b) For cylinders the intercept is 1/16 = 0.0625. Thus the answer must be

$$D_{\text{p, zero collection}} = 7.4\ \mu\sqrt{\frac{0.0625}{3.4}} = 1.003\ \mu$$

9.56 (a)

$$N_s = \frac{\left(2000\dfrac{\text{kg}}{\text{m}^3}\right)\cdot(1\,\text{E-}5\text{ m})\cdot\left(1\dfrac{\text{m}}{\text{s}}\right)}{18\cdot\left(1.8\,\text{E-}5\dfrac{\text{kg}}{\text{m}\cdot\text{s}}\right)\cdot(300\,\text{E-}6\text{ m})} = 2.06, \quad \eta_t = 0.72$$

$$\eta = 0.72\cdot 0.75 = 0.54 = 54\%$$

At the cut diameter, $\eta = 0.5 = \dfrac{\eta_t}{0.75}, \quad \eta_t = \dfrac{0.5}{0.75} = 0.666, \quad N_s \cong 1.6$

$$\frac{D_c}{10\ \mu} = \sqrt{\frac{N_s}{2.06}}; \quad D_c = 10\ \mu\sqrt{\frac{1.6}{2.06}} = 8.8\ \mu \cong 9\ \mu$$

(b) $p_2 = (p_1)^2 = (1-0.54)^2 = 0.2116; \quad \eta_2 = 0.79 = 79\%$

(c) $p_{\text{i, cut}} = \sqrt{p_{\text{overall, cut}}} = \sqrt{0.5} = 0.707; \quad \eta_{\text{t,cut}} = 1 - 0.707 = 0.293$

$$\eta_{\text{t,cut}} = \frac{0.293}{0.75} = 0.390, \quad N_s = 0.52; \quad D_c = 10\ \mu\sqrt{\frac{0.52}{2.06}} = 5.0\ \mu$$

(d) For $D = 2.5\mu$, $N_s = 2.06\left(\dfrac{2.5}{10}\right)^2 = 0.1287, \quad \eta_t = 0.04, \quad \eta = 0.04\cdot 0.75 = 0.03$

$$p_{\text{overall, cut}} = 0.5 = (1-0.03)^n; \quad n = 22.8$$

I actually measured a handkerchief on an optical comparator; the numbers in this problem correspond to those measurements. Industrial dust masks are the equivalent of this problem, generally with finer fibers in felted form. The filters in those masks have much finer fibers, and can collect particles down to about 0.3 μ. They operate at much lower velocities (much higher filtering areas) than the simple bandana shown here, and collect the smallest particles largely by diffusion.

The Lone Ranger's mask was not of this type, but was a domino, which covers the eyes. Tonto wore a bandana over his nose and mouth, as in this problem.

9.57 Rewriting Equation 9.46 we have

$$\eta_t = \frac{6\left(\frac{\gamma}{D}\right)^{\frac{2}{3}}}{\nu^{\frac{1}{6}} D_b^{\frac{1}{2}} V^{\frac{1}{2}}} + \frac{3D^2 V^{\frac{1}{2}}}{\nu^{\frac{1}{2}} D_b^{\frac{2}{3}}} = \frac{\alpha}{D^{\frac{2}{3}}} + \beta D^2$$

where

$$\alpha = \frac{6\gamma^{\frac{2}{3}}}{\left(\nu^{\frac{1}{6}} D_b^{\frac{1}{2}} V^{\frac{1}{2}}\right)} \text{ and } \beta = \frac{3V^{\frac{1}{2}}}{\left(\nu^{\frac{1}{2}} D_b^{\frac{2}{3}}\right)}$$

Then

$$\frac{d\eta_t}{dD} = -\frac{2}{3}\frac{\alpha}{D^{\frac{5}{2}}} + 2\beta D; \quad \text{set} = 0, \quad D^{\frac{8}{3}} = \frac{\alpha}{3\beta} = \frac{2\gamma^{\frac{2}{3}} D_b \nu^{\frac{1}{3}}}{3V}, \quad \text{etc.}$$

In the original article the constant is shown as 0.885, while $(2/3)^{3/8}$= 0.889. Presumably in 1958 the 3/8 root was extracted with a slide rule, with the result shown.

For the conditions in Example 9.21, the diameter with minimum collection efficiency is

$$\gamma = \frac{kT}{3\pi\mu} = \frac{1.38\text{ E-23}\cdot\left(\frac{\text{kg}\cdot\text{m}^2}{\text{s}^2\cdot\text{K}}\right)\cdot(293.15\text{ K})}{3\pi\left(1.8\text{ E-5}\frac{\text{kg}}{\text{m}\cdot\text{s}}\right)} = 2.38\text{ E-17}\frac{\text{m}^3}{\text{s}}$$

$$D_{\min} = \frac{0.859\cdot\left(2.38\text{ E-17}\frac{\text{m}^3}{\text{s}}\right)^{\frac{1}{4}}\cdot\left(1.49\text{ E-5}\frac{\text{m}^2}{\text{s}}\right)^{\frac{1}{8}}\cdot(1\text{ E5 m})^{\frac{3}{8}}}{\left(1\frac{\text{m}}{\text{s}}\right)^{\frac{3}{8}}} = 1.99\text{ E-7 m} \cong 0.2\mu$$

One may compute that for this diameter, and the conditions of Example 9.22 the target efficiencies are: impaction, 0; diffusion, 0.0054; interception, 0.0005.

---

9. 58 (a) C = 1.24

$$p = \exp-\left[\frac{(7)\cdot(7\text{ft})\cdot\left(0.2\frac{\text{ft}}{\text{s}}\right)\cdot(0.5\text{ E-6 m})^2\cdot\left(2000\frac{\text{kg}}{\text{m}^3}\right)\cdot(1.24)}{9\cdot(0.001\text{ m})^2\cdot\left(1.8\text{ E-5}\frac{\text{kg}}{\text{m}\cdot\text{s}}\right)\cdot(0.3)}\cdot\left(\frac{\text{m}}{3.28\text{ ft}}\right)^2\right]$$

$$= \exp(-11.62) = 8.98\text{ E-6}, \quad \eta = 99.9991\%$$

(b)

$$\text{For } p = 0.001 = \exp(-\alpha); \quad \alpha = 6.90, \quad z = 7\text{ft}\cdot\frac{6.90}{11.62} = 4.16\text{ ft} = 1.27\text{ m}$$

This was a wartime expedient, used while we thought we might be losing the race with the Germans to develop the atom bomb. The filters were like swimming pools, filled with sand, with a distributor pipe network at the bottom. When the radioactive particles began to appear at the top, the pools were paved over, and left for future generations to deal with. In 1999 we are beginning to deal with them. One would not do this now, but there are similar filters, in which the dirty sand is cleaned and reused.

One may compute that for this flow rate the expected pressure drop across the 7 ft bed is about 1 psi, which is about as much as one would normally use.

9.59 Start with Equation 9.76, in penetration form. Substitute:

$$N = \frac{1}{2}\frac{z}{D_c}; \text{ i.e.} \frac{1}{2} \text{turn per particle} ; \ .w_i \cong 0.25 D_c, \quad V = \frac{V_s}{\varepsilon}$$

$$p = \exp - \left\{ \frac{\frac{\pi}{9} \cdot \frac{1}{2} z \left(D_{pa} \right)^2 V_s}{D_c \cdot 0.25 D_c \cdot \mu \cdot \varepsilon} \right\};$$

If 2 = 7 then this is Eq. 9.76. 2 = 6.28.

One can also rewrite Equation 9.19 as

$$p = \exp - (2\pi N) \cdot \left(\frac{V_c D_p{}^2 \rho_p}{18 w_i \mu} \right) = \exp(-2\pi N) \cdot N_s$$

which shows that N_S plays a fundamental role in any kind of particle collection device which used the particle's inertia, which a cyclone does.

One student asked me in class why I used the Cunningham correction factor in Problem 9.50 while it does not appear in the standard equations for cyclones or this problem. I asked the class to answer the question. After some thought a few of them saw that for the size particle normally collected in cyclones, the CCF is 1, and is ignored. In the most careful work we would put it in. For the very small cyclones used in analytical devices one would have to take it into account. In this type of filter, the particles are small enough that it is normally taken into account.

9.60 (a) From Figure 8.7, V_t 22 ft/s 7 m/s, so that

$$N_s = \frac{(3\,\text{E-6 m})^2 \cdot \left(7 \frac{\text{m}}{\text{s}}\right) \cdot \left(2000 \frac{\text{kg}}{\text{m}^3}\right)}{18 \cdot \left(1.8\ \text{E-5} \frac{\text{kg}}{\text{m} \cdot \text{s}}\right) \cdot (0.002\ \text{m})} = 0.194; \quad \eta_t = 0.18$$

$$\text{Volume Swept} = \frac{\pi}{4} D^2 L = \frac{\pi}{4} \cdot (0.002\ \text{m})^2 \cdot \left(\frac{1000\ \text{m}}{3.28}\right) = 9.58\ \text{E-4 m}^3$$

$$\text{Mass Captured} = Vc\eta_t = \left(9.58\ \text{E-4 m}^3\right) \cdot \left(80\ \text{E-6} \frac{\text{g}}{\text{m}^3}\right) \cdot (0.18) = 1.38\ \text{E-8 g}$$

$$\text{Mass per particle} = \frac{\pi}{6}D^3\rho = \frac{\pi}{6}\cdot(3\,\text{E-6 m})^3\cdot\left(2000\frac{\text{kg}}{\text{m}^3}\right) = 2.83\,\text{E-14 kg} = 0.014\ \mu\text{g}$$

$$\text{Number of particles} = \frac{m_T}{m_p} = \frac{1.38\,\text{E-8 g}}{2.83\,\text{E-14kg}}\cdot\frac{\text{kg}}{1\,\text{E6 g}} = 490\ \text{particles}$$

(b) the mass of the drop is

$$m_{\text{drop}} = \frac{\pi}{6}D^3\rho = \frac{\pi}{6}\cdot(2000\ \text{m})^3\cdot\left(1000\frac{\text{kg}}{\text{m}^3}\right) = 4.18\,\text{E-6 kg} = 0.00418\ \text{g}$$

so that

$$\frac{\Delta m}{m} = \frac{1.38\,\text{E-8 g}}{4.18\,\text{E-3 g}} = 3.29\,\text{E-6}$$

9.61** Starting from Equation 9.57,

$$\frac{\text{Drops}}{\text{volume}} = \frac{N_D\Delta z}{V_t - 0}\cdot\frac{1}{\Delta x\Delta y\Delta z};\quad N_D = \frac{Q}{\frac{\pi}{6}D^3}$$

$$\frac{\text{Drops}}{\text{volume}} = \frac{Q}{\Delta x\Delta y}\cdot\frac{1}{\frac{\pi}{6}D^3V_t} = \frac{Q}{A}\cdot\frac{1}{\frac{\pi}{6}D^3V_t}$$

From Figure 8.7, V_t 22 ft/s, so that

$$\frac{\text{Drops}}{\text{volume}} = \frac{1\ \text{ft}}{12\ \text{hr}}\cdot\frac{6}{\pi}\cdot\frac{1}{(0.002\,\text{m})^3}\cdot\frac{\text{s}}{22\,\text{ft}}\cdot\frac{\text{hr}}{3600\,\text{s}} = 251\frac{\text{drops}}{\text{m}^3} = 7.1\frac{\text{drops}}{\text{ft}^3}$$

9.62 See Example 9.22. Starting with Eq. 9.52 and rearranging

$$\Delta t = \frac{(-\ln p)(D_0)}{1.5\eta_t\left(\frac{Q_L}{A}\right)} = \frac{(-\ln 0.1)\cdot(0.002\,\text{m})}{(1.5)\cdot(0.1)\cdot\left(0.01\frac{\text{in}}{\text{hr}}\right)}\cdot\frac{\text{in}}{0.0254\,\text{m}} = 12.09\,\text{hr}$$

I ask the students whether the D in this formula was the particle diameter or the drop diameter. After a few minutes pause about half of them reply that it is the drop diameter. Then I ask them why the particle diameter does not appear. After some head scratching one or more of them will figure out that it does appear, but hidden in the value of the target efficiency.

9.63 See Example 9.22. One can solve by assuming that all the large drops fall, and then all the small drops fall. The overall penetration is the product of the penetrations for each of the drops falling separately. It makes no difference whether large fall first or last. To prove this write the Equation 9.51 for *dc/c* with two terms on the right, one for each size, integrate and rearrange to show that the overall penetration is the product of the two individual penetrations. For each size Q_L = 0.5 mm/min = 12.7 mm/hr. Then

$$p = \exp\left(-\frac{1.5\eta_t Q_L \Delta t}{D_D A}\right)_{\text{large}} \cdot \exp\left(-\frac{1.5\eta_t Q_L \Delta t}{D_D A}\right)_{\text{small}}$$

$$= \exp\left(\frac{-1.5\cdot(0.1)\cdot\frac{12.7\text{mm}}{\text{hr}}\cdot 1\text{hr}}{2\text{ mm}}\right)\cdot \exp\left(\frac{-1.5\cdot(0.15)\cdot\frac{12.7\text{mm}}{\text{hr}}\cdot 1\text{hr}}{1\text{mm}}\right)$$

$$= 0.3857\cdot 0.05741 = 0.0221;\quad \eta = 1 - 0.0221 = 0.9778 = 97.8\%$$

9.64 See the solution to the previous problem.

(a) $p = \exp\left(-\frac{1.5\cdot(0.1)\cdot(0.9\text{inch})}{2\text{mm}}\cdot\frac{25.4\text{mm}}{\text{in}}\right)\cdot\exp\left(-\frac{(1.5)\cdot(0.3)\cdot(0.1\text{inch})}{0.5\text{mm}}\cdot\frac{25.4\text{mm}}{\text{in}}\right)$

$= (0.1016)\cdot(0.180);\quad \eta = 0.9817$

(b) Here we write Equation 9.51 twice, one for the small drops, one for the large, and divide one by the other, finding

$$\frac{\left(\frac{-dc}{dt}\right)_{\text{small}}}{\left(\frac{-dc}{dt}\right)_{\text{large}}} = \frac{c}{c}\cdot\frac{\frac{\eta_S}{D_S}\left(\frac{Q_L}{A}\right)_S}{\frac{\eta_L}{D_L}\left(\frac{Q_L}{A}\right)_L} = \frac{\frac{0.3}{0.5\text{mm}}\cdot(0.1\text{inch})}{\frac{0.1}{2\text{mm}}(0.9\text{inch})} = 1.333 = \frac{m_s}{m_L}$$

From which

$$m_T = m_L + m_S;\quad \frac{m_T}{m_S} = \frac{m_L}{m_S} + 1 = \frac{1}{1.333} = 1.750$$

$$\frac{m_S}{m_T} = \frac{1}{1.750} = 0.571;\quad \frac{m_L}{m_T} = 1 - 0.571 = 0.429$$

9.65 Writing Equation 9.54 twice, dividing one by the other, we find

$$\frac{\ln p_{\text{new}}}{\ln p_{\text{old}}} = \frac{Q_{G_{\text{old}}}}{Q_{G_{\text{new}}}} = \frac{1}{2};\quad \ln p_{\text{new}} = \frac{1}{2}\ln(0.10) = -1.1513$$

$$p_{\text{new}} = \exp(-1.1513) = 0.316;\quad \eta_{\text{new}} = 0.684 = 68.4\%$$

9.66 Writing Equation 9.54 twice, dividing one by the other, we find

$\frac{\ln p_{\text{new}}}{\ln p_{\text{old}}} = \frac{\eta_{t_{\text{new}}}}{\eta_{t_{\text{old}}}}\cdot\frac{D_{\text{old}}}{D_{\text{new}}}$ Here from Figure 8.7 we find that the two sizes of drops have terminal velocities of 5.2 and 2.4 ft/s. For the larger drops the separation number is

$$N_s = \frac{(3\,\text{E-6 m})^2 \cdot \left(5.2\frac{\text{ft}}{\text{s}}\right) \cdot \left(2000\frac{\text{kg}}{\text{m}^3}\right)}{18 \cdot \left(1.8\,\text{E-5}\frac{\text{kg}}{\text{m}\cdot\text{s}}\right) \cdot (4\,\text{E-4 m})} \cdot \frac{0.304\text{ m}}{\text{ft}} = 0.220; \quad \eta_t \approx 0.22$$

and for the smaller

$$N_s = 0.220 \cdot \left(\frac{2.4}{5.2}\right) \cdot \left(\frac{400}{200}\right) = 0.203; \quad \eta_t \approx 0.20$$

so that

$$\ln p_{\text{new}} = \ln(0.10) \cdot \frac{0.20}{0.22} \cdot \frac{400}{200} = -4.187$$

$$p_{\text{new}} = \exp(-4.187) = 0.015; \quad \eta = 0.985 = 98.5\%$$

Here in the discussion, make clear that we will have a problem with the smaller drops being carried out the downstream end of the scrubber. An additional interesting discussion item here is the fact that at this drop size range the terminal velocity is practically proportional to the drop diameter. For small drops velocity is proportional to the square of the diameter, for large drops to the square root of the diameter. This size is near the "knee" of the curves on Figure 8.7, where the slope is practically unity.

9.67** $\ln p \propto Q_L; \quad \frac{Q_{L_{new}}}{Q_{L_{old}}} = \frac{\ln(0.05)}{\ln(0.10)} = 1.30$

The liquid flow rate must be increased by 30%.

9.68 Here $\frac{Q_L}{Q_G} = \frac{10}{10^4} = 0.001,$

$$V_G = \frac{Q_G}{A} = \frac{1\,\text{E4}\frac{\text{ft}^3}{\text{min}}}{0.6\,\text{ft}^2} = 1667\frac{\text{ft}}{\text{min}} = 278\frac{\text{ft}}{\text{s}} = 84.7\frac{\text{m}}{\text{s}}$$

(a) From Figure 9.27 we read $\Delta P \approx 70\,\text{cm H}_2\text{O}, \quad D_{cA} \approx 0.8\,\mu$ aerodynamic

We can check the pressure drop from Equation 9.67

$$P_1 - P_2 = \frac{Q_L Q_G \rho_L}{A} = \frac{10\frac{\text{ft}^3}{\text{min}} \cdot 1\,\text{E4}\frac{\text{ft}^3}{\text{min}} \cdot \frac{62.3\,\text{lbm}}{\text{ft}^3}}{0.6\,\text{ft}^2} \cdot \frac{\text{lbf s}^2}{32.2\,\text{lbm ft}} \cdot \left(\frac{\text{min}}{60\,\text{s}}\right)^2$$

$$= 149.5\frac{\text{lbf}}{\text{ft}^2} = 1.04\,\text{psi} = 2.39\text{ ft H}_2\text{O} = 73\text{ cm H}_2\text{O}$$

(b) To find the physical particle diameter we rearrange the definition of the aerodynamic cut diameter, and then assume that $C = 1$, so that

$$D_{\text{physical}} = \frac{D_{\text{aerodynamic}}}{\sqrt{\rho\left(\frac{\text{gm}}{\text{cm}^3}\right)}} = \frac{0.8\ \mu\text{a}}{\sqrt{2}} = 0.57\ \mu$$

If we retain the Cunningham correction factor, that leads to $D\sqrt{1+A\frac{\lambda}{D}} = \frac{D_{aero}}{\sqrt{\rho}}$ if we square both sides we find a quadratic equation for D, with solution D 0.5 μ

9.69 $\sigma_{\text{geometric}} = \exp\sigma = \exp 1.5 = 4.48$ and $D_{\text{cut, aero}} / D_{\text{mean, aero}} \approx D_{\text{cut}} / D_{\text{mean}}$. This is shown as because while the densities are the same, the C's are not. For this problem $D_{\text{cut}} / D_{\text{mean}} = 0.5\ \mu / 2.0\ \mu = 0.25$. By chart lookup on Fig 9.30 $p \approx 0.18$; $\eta = 0.82$.

Chapter 10

10.1** (a) CO_2 is the only substance shown whose triple point is greater than 0.1 psi For it the leftmost line is the vapor-solid equilibrium, the rightmost the vapor-liquid equilibrium. Liquid cannot exist below 75 psia. UF_6 also has a triple point above 1 atm, and perhaps some other materials, but it is rare to have that high a triple point pressure.

(b) For water the triple point pressure is 0.08865 psia, so that if the plot were extended down only slightly from 0.1 psia, the triple point of water would be shown.

10.2** (a) Using the constants from Appendix A

$$\log_{10} p = 7.975756 - \frac{3255.61}{(20 + 281.988)} = -2.8048,$$

$$p = 0.001567\,\text{mm Hg} = 2.06\text{ E - 6 atm}$$

This is very close to the answer in Example 10.3. As discussed in (b) below, if $C =$ 273.15 the answers should be identical, here $C = 281.98$, which is quite close to 273.15, so we would expect that result.

(b) $C = 273.15$. Most of the values in Appendix A are in the range 200 to 240.

(c) For $C= 273.15$ it is also a straight line. For $C< 273$, which is the most common value in Appendix A, it is curve with a positive second derivative (i.e. concave upward).

10.3 The steam table values for 0.1, 0.3, 1,...1000, 3203.6 psi can be curve fit, with $R^2 = 0.999$ by $\log_{10} p(\text{psia}) = 6.8759 - 3.8649 \cdot 10^3/T(°\text{R})$. Then, using a spreadsheet, one can compute the ratio of the pressure computed from this equation to the steam table value, over the same range, finding that the values form an orderly progression, starting at 0.992, increasing to 1.008, and then declining to 0.985.

The point of this problem is for the students to learn that the simple Clausius Clapyron equation is good but not perfect. The "Cox Chart" approach was widely used in the age of the slide rule, but is much less used in the age of the computer. Mostly it has been replaced by the Antoine equation and tables like those in Appendix A, which are all convenient for computer calculations.

Using the Antoine equation with $C = 43.15$ one may solve for the other constants, finding $\log_{10} p = 6.2876 - \dfrac{1689.4}{(T(\text{K}) - 43.15)}$ This correlates the above listed data with $R =$ 0.99999. On a plot one can see no deviation from linearity. However if one computes the pressures from this equation and takes their ratios to the reported pressures one sees errors up to 1.4 %, plus a 3.7% error at the critical point. So it is not a perfect straight line on these coordinates.

10.4 (a) Ingestion $= \dfrac{15\,\text{kg}}{\text{day}} \cdot \dfrac{\text{m}^3}{1.20\,\text{kg}} \cdot \dfrac{\text{day}}{3\,\text{shifts}} \cdot \dfrac{0.025\text{mg}}{\text{m}^3} = 0.104 \dfrac{\text{mg}}{\text{shift}}$

If that were condensed to a drop of liquid mercury, it would have a diameter of about 376 μ and be plainly visible.

(b) Basis; 1 m^3 of air. Mols air = $41.6 \frac{\text{mol}}{\text{m}^3}$

$$\text{Mols mercury} = \frac{(0.025 \cdot \text{E} - 3\ \text{g})}{\left(200.6 \frac{\text{g}}{\text{mol}}\right)} = 1.25\ \text{E - 7} \frac{\text{mol}}{\text{m}^3}$$

$$\text{Mercury mol fraction} = \frac{1.25\ \text{E - 7}}{41.6} = 3.00\ \text{E - 9} \cong 3\ \text{ppb}$$

(c) Mercury vapor pressure $1.6 \cdot 10^{-6}$ atm, so that at saturation the mole fraction of mercury would be 1600 ppb, 533 times the TLV.

(d) Equating the mercury flux from the cracks to the mercury leaving the room in the ventilation air (this corresponds to simple box model in Sections 6.2 and 15.4.2.1)

$$A = \frac{Qc}{F} = \frac{\left(75\text{m}^3 \cdot \frac{2}{\text{hr}}\right)\left(1.25\ \text{E - 7} \frac{\text{mol}}{\text{m}^3}\right)}{6\ \text{E} - 10 \frac{\text{mol}}{\text{cm}^2 \cdot \text{s}}} \cdot \frac{\text{hr}}{3600\,\text{s}} = 8.7\,\text{cm}^2$$

For cracks 1 mm wide, this corresponds to 8.7 m 2.8 ft of cracks. That seems like a large number. I recommend rapid and careful cleanup of all mercury spills, but it does not appear that small ones pose a severe immediate hazard to life and health.

This is not strictly an air pollution problem, but I included it because it ties in with the subject matter here, is of some industrial hygiene significance, and ties in with the personal experience of the students.

In the "University Folklore" there is a well known story of a university chemistry department which decided that some of its old alums must have made a lot of money, and thus be suitable subjects for an approach for a large gift. They found that they had no old alums, presumably because they had all been exposed to so much mercury as undergraduates that they didn't live to be old alums. Based on this calculation I think that part of the folklore is probably not factual. It is certainly true that when I was an undergraduate in the chemistry laboratories at Stanford in 1950-54, the floors were wooden, old, and had lots of cracks between boards. But I remember that we cleaned up mercury spills pretty quickly and thoroughly. Currently we are very careful with mercury in our laboratories, and, for example, only use mercury manometers if the user can show that there is no other practical alternative.

10.5** In general have $\frac{dm}{dt} = \rho\frac{dV}{dt} = \rho A\frac{dx}{dt} = -A\alpha M\frac{p}{P}$ where α is the coefficient in the evaporation equation. Solving this for dt and then integrating from the initial thickness, x_o to 0, we find $\Delta t = \frac{\rho x_o}{\alpha M\frac{p}{P}}$

(a)

$$\Delta t = \frac{(1\text{ E-}3\text{m})\cdot\left(1000\frac{\text{kg}}{\text{m}^3}\right)}{\left(0.5\frac{\text{mol}}{\text{m}^2\cdot\text{s}}\right)\cdot\left(0.4\frac{\text{kg}}{\text{mol}}\right)\cdot\left(\frac{1\text{ E-}7\text{torr}}{760\text{torr}}\right)} = 3.8\text{ E10 s} = 1205\text{years}$$

(b)

$$\Delta t = \frac{(1\text{ E-}3)\cdot(753)}{(0.5)\cdot(0.060)\cdot\left(\frac{6}{14.7}\right)} = 61\text{ s}$$

(c) Here I have linearized the two curves on Fig 10.3. The true situation is more complex, but this is approximately right. Replacing the x in the general solution with dx, we have

$$\int dt = \int\frac{P}{p}\frac{\rho}{\alpha M}(-dx)$$

then we let $-dx = x_0 dy$ where x_o is the original thickness, and y is the fraction evaporated. This changes the two equations in the problem to

$$p = p_0(1-0.5y) \quad\text{and}\quad M = M_0(1+y)$$

$$\int dt = \int_{y=0}^{y=1}\frac{P}{p_0(1-0.5y)}\cdot\frac{\rho}{\alpha}\cdot\frac{1}{M_0(1+y)}\cdot x dy = \frac{P\rho x_0}{p_0\alpha M_0}\int_0^1\frac{dy}{(1-0.5y)(1+y)}$$

$$= \begin{pmatrix}\text{solution}\\ \text{to part (b)}\end{pmatrix}\cdot\left|\frac{1}{(1)(1)-(-0.5)(1)}\ln\left(\frac{1+y}{1-0.5y}\right)\right|_{y=0}^{y=1}$$

$$= 61.25\text{ s}\cdot\frac{1}{1.5}\cdot\left(\ln\frac{2}{0.5} - \ln 1\right) = 61.25\text{ s}\cdot 0.924 = 56.6\text{ s}$$

This gives the surprising result that even with the declining vapor pressure the evaporation time does not decrease. The reason is that the evaporation time is proportional to the average value of the product of the p and M. For the simple relations assumed here that product is always equal to or a bit larger than the value in part (b). In practice the tails on Figure 10.3 make the final part of the evaporation slower than the value calculated here. Nonetheless one can try this, pouring a mm of gasoline into a shallow-rimmed dish (outdoors, away from ignition sources) and find that the evaporation time is of the order of one minute. See Problem 10.6 for more on the mass transfer coefficient used in this problem.

10.6 (a) For these conditions $G_m = 10\frac{\text{ft}}{\text{s}} \cdot 0.75\frac{\text{lbm}}{\text{ft}^3} \cdot \frac{\text{lb mol}}{29\,\text{lbm}} = 0.0259\frac{\text{lb mol}}{\text{ft}^2 \cdot \text{s}}$

$$k'_g = \frac{j_m G_m}{(S_c)^{\frac{2}{3}}} = \frac{(2.7\,\text{E-3}) \cdot \left(0.0259\frac{\text{lb mol}}{\text{ft}^2 \cdot \text{s}}\right)}{(1)^{\frac{2}{3}}} = 7.0\,\text{E-5}\frac{\text{lb mol}}{\text{ft}^2 \cdot \text{s}} = 0.34\frac{\text{mol}}{\text{m}^2 \cdot \text{s}}$$

In Problem 10.5 the constant is shown as 0.5, same units. Thus the expression in that problem is the (0.5/0.34 1.4) times the value shown here. The simple mass transfer relation shown in Problem 10.5 would be expected to be close to right for typical outdoor wind velocities (See Chapter 5).

(b) Comparing their equation to the one in Problem 10.5 we find that

$$N = \dot{n} = \frac{k_m}{RT}p = \alpha\frac{p}{P}; \quad \alpha = \frac{k_m}{RT} \quad (\text{for } p \text{ in atmospheres})$$

Then, using the values from the problem, which must be changed to the units stated, we find

$$\alpha = \frac{(0.0292) \cdot U^{0.78} X^{-0.11} S_c^{0.67}}{RT} = \frac{(0.0292) \cdot (10{,}800)^{0.78} (2)^{-0.11} (2)^{0.67}}{8.206\,\text{E}-5 \cdot 298}$$

$$= 973\frac{\text{mol}}{\text{m}^2 \cdot \text{hr}} = 0.27\frac{\text{mol}}{\text{m}^2 \cdot \text{s}}$$

which is about half of the value given in Prob. 10.5. Based on the finding in these two problem I suspect that my value of α in Prob. 10.5 may be too high.

10.7 (a) Basis: 1 L of paint, $\rho = 1\frac{\text{g}}{\text{cm}^3}$: 0.6 kg solvent; , 0.4 kg solids

$$\frac{\text{solids}}{\text{solvent}} = \frac{0.6}{0.4} = 1.5\frac{\text{kg}}{\text{kg}}; \quad p = \frac{0.29\frac{\text{kg}}{\text{kg}}}{1.5\frac{\text{kg}}{\text{kg}}} = 0.193; \quad \eta = 0.807 = 80.7\%$$

(b) Either use a water based paint, which has much less (but not zero) VOC per weight of solid, or go to a very high solids content paint. The latter is probably not practical for paints used outdoors, so that, as far as I know, this regulation has the practical effect of banning oil based paints for outdoor usage.

The Glidden Paint Company (ICI) has introduced a water emulsion paint which emits no VOC on drying. They have developed a polymer system in which the small amount of solvent (about 7% by volume of the total paint mix) which is used in conventional latex paints to allow the polymer droplets to fuse properly to make a good paint film is not needed.

10.8 (a) by direct substitution

$$D_L = 12.46 \cdot \left[\frac{1 \cdot 1.45 \cdot 78}{528} \right] = 2.67 \frac{\text{lb}}{1000 \text{ gal}}$$

which is identical to the result in Example 10.4, as it must be.

(b) In submerged loading, the liquid entry pipe is below the liquid level in the tank, in splash loading it is above the liquid level, and the liquid entering the tank falls and splashes on the liquid surface. We fill our coffee cups, bathtubs and the gasoline tanks of our private autos by splash loading. If the level in the bathtub rose above the level of the fill spout, then we would fill the bathtub by submerged loading. Most modern gasoline tankers, which deliver gasoline to service stations have submerged loading; all the piping is on the bottom of the tank.

With submerged loading there is little mixing between the liquid entering the tank and the gas being expelled. With splash loading not only will the expelled gas be saturated with the liquid being loaded, but it will probably contain droplets of the liquid, formed by the splashing at the liquid surface (hence $S > 1$). If time permits one can discuss the ingenious connectors and valves which have been developed for this kind of liquid transfer. Many of them allow lines up to 3 inches in diameter to be used, and then disconnected, with only a teaspoon of liquid lost in the process!

In Section 4.8 I note that some emission factors are calculable (generally from stoichiometry) while others are not, and must be based entirely on emission measurements. This is an example (like Example 4.4) of an emission factor which can be calculated, but which contains a coefficient (in this case S), which is based on experimental measurements.

10.9 From Prob. 10.4 we know that at saturation the mol fraction of mercury in the headspace is 1.6 E-6. The 1.00 m^3 of headspace gases expelled corresponds to 41.6 mol, so that

$$\begin{pmatrix} \text{mass of Hg} \\ \text{expelled} \end{pmatrix} = 41.6 \text{ mol} \cdot 1.6 \text{ E-6} \cdot \frac{200.6 \text{ g}}{\text{mol}} = 0.0133 \text{ g} = 2.9 \text{ E-5 lb}$$

10.10** From Example 10.7 the displacement losses are 0.00134 of the liquid added. The permitted tailpipe emissions of HC are

$$\frac{\text{tailpipe emissions}}{\text{gasoline filled}} = \frac{0.41 \frac{\text{g}}{\text{mi}}}{\left(\frac{\text{gal}}{20 \text{ mi}}\right) \cdot \left(\frac{\text{ft}^3}{7.48 \text{ gal}}\right) \cdot \left(0.753 \cdot 62.4 \frac{\text{lb}}{\text{ft}^3}\right) \cdot \left(454 \frac{\text{g}}{\text{lb}}\right)} = 2.87 \text{ E} - 3$$

and the ratio is

$$\frac{(\text{displacement loss})}{(\text{tailpipe emissions})} = \frac{1.34 \text{ E} - 3}{2.87 \text{ E} - 3} = 0.466 = 47\%$$

This is a very significant part of the overall HC emissions, and shows why either Stage II or enlarged on-board canisters are probably a good idea. As discussed in the text, the decision has been made in favor of increased size onboard canisters, and closed filling of gasoline tanks, which began with some 1999 models and must be completed in a few years.

10.11 See Example 10.5. The first part is the same $\Delta V/V = 0.039$. For the second part the vapor pressures are $p_{70° F} = 6.32 \text{ psia}, \quad p_{100° F} = 10.73 \text{ psia}$ and

$$\frac{\Delta V}{V} = 0.5 \cdot \left(\frac{10.73 - 6.32}{14.7} \right) = 0.153; \quad \left(\frac{\Delta V}{V} \right)_{\text{total}} = 0.039 + 0.153 = 0.192$$

$$c_{\text{gasoline vapor, initial}} = \frac{6.32}{14.7} \cdot \frac{60}{29} \cdot 0.075 \frac{\text{lb}}{\text{ft}^3} = 0.0667 \frac{\text{lb}}{\text{ft}^3}$$

$$\left(\frac{\text{gasoline expelled}}{\text{tank volume}} \right) = 0.192 \cdot 0.0667 = 0.0128 \frac{\text{lb gasoline}}{\text{ft}^3 \text{of tank}}$$

$$\text{@ 50\% liquid full,} \quad \frac{m_{\text{gasoline}}}{V_{\text{tank}}} = 0.5 \cdot 0.72 \cdot 62.4 \frac{\text{lb}}{\text{ft}^3} = 22.4 \frac{\text{lb}}{\text{ft}^3}$$

$$\text{Fraction expelled} = \frac{0.0128}{22.4} = 5.7 \text{ E} - 4$$

10.12 Here 1% corresponds to a change of 0.15 psia = 0.29 in Hg = 10 mb. This is about 1/3 of the maximum change in atmospheric pressure between a high and low pressure system, See Chapter 5. Then see Example 10.5. We rewrite Eq. 10.10 as follows

$$\frac{dV}{V} = \left(\frac{V_{\text{vapor}}}{V_{\text{tank}}} \beta_{\text{vapor}} + \frac{V_{\text{liquid}}}{V_{\text{tank}}} \beta_{\text{liquid}} - \beta_{\text{tank}} \right) dP$$

where β is the cubical expansion coefficient (= 3 times the Young's modulus). The value is assumed negligible for the tank itself. For a perfect gas

$$\beta = \frac{\left(\frac{dV}{V} \right)}{dP} = \frac{1}{V} \cdot \frac{dV}{dT} = \frac{P}{RT} \cdot \frac{d}{dP} \left(\frac{RT}{P} \right) = \frac{-1}{P}$$

so that

$$\frac{dV}{V} = \left(\frac{V_V}{V_T} \cdot \frac{dP}{(-P)} + \frac{V_L}{V_T} \beta_L dP \right)$$

$$= (0.5) \cdot (0.01) + 0.5 \cdot \left(\frac{7 \text{ E-6}}{\text{psi}} \right) \cdot (0.01 \cdot 14.7 \text{ psia}) = 0.005 + 5 \text{ E} - 7 \approx 0.005$$

The term corresponding to the second step of that example is

$$\frac{\Delta V}{V} = \frac{V_V}{V_T} \cdot \left(\frac{p_i}{P_{\text{final}}} - \frac{p_i}{P_{\text{initial}}} \right) = 0.5 \cdot \frac{6\,\text{psi}}{14.7\,\text{psi}} \cdot \left(\frac{1}{0.99} - \frac{1}{1} \right) = 0.00196 \approx 0.002$$

and the total fraction of the tank volume expelled is

$$\frac{\Delta V}{V} = 0.005 + 0.002 = 0.007$$

This is much less than the other losses described in that section of the book. For that reason vapor losses due to atmospheric pressure changes are normally ignored, except in the most detailed analyses.

10.13 (a) $\quad \text{Fraction} = 0.32 \frac{\text{g}}{\text{gal}} \cdot \frac{7.48\,\text{gal}}{\text{ft}^3} \cdot \frac{\text{ft}^3}{47\ \text{lbm}} \cdot \frac{\text{lbm}}{454\ \text{g}} = 1.12\ \text{E} - 4$

(b) For 12 gallons, the loss is

$$\text{Loss} = 12\ \text{gal} \cdot 1.12\ \text{E - 4} = 0.00135\ \text{gal} = 0.31\ \text{in}^3 = 5.10\ \text{cm}^3 = 1.03\ \text{teaspoons}$$

This seems much larger than my observation with my car; the automatic shutoffs work better than that. However when I asked this question of some oil company people I was told that many consumers and some service station attendants "top off" the customer's tank, by manually overriding the automatic shutoff, thus greatly increasing the spillage.

(c)

$$\text{Combined Loss} = 0.45 \frac{\text{g}}{\text{mile}} \cdot \frac{12.7\ \text{mile}}{\text{gal}} \cdot \frac{7.48\ \text{gal}}{47\ \text{lbm}} \cdot \frac{\text{lbm}}{454\,\text{g}} = 0.0020$$

$$\text{Spillage} = \text{Combined - Displacement} = 0.0020 - 0.00134 \begin{pmatrix} \text{See} \\ \text{Ex.10.7} \end{pmatrix} = 0.00066$$

This is five times the value computed in (a). It is hard to get good data on the average spillage loss in actual practice; this disagreement should not surprise us.

10.14** The flow rate is proportional to the third power of the thickness of the leakage path, so that

$$\frac{h_2}{h_1} = \left(\frac{Q_2}{Q_1} \right)^{\frac{1}{3}} = \left(\frac{0.024}{0.007} \right)^{\frac{1}{3}} = 1.508; \quad h_2 = 1\ \text{E - 4 in.} \cdot 1.508 = 1.508\ \text{E - 4 in.}$$

True leakage paths are certainly more complex than the uniform annulus assumed here, but this description of the leakage path is generally correct.

In that example I treated the annular space between the stem and the packing as if it were a rectangular slit. You might ask bright students how much error is introduced that way? I checked by comparing the slit solution to that for an annulus with inside diameter 0.25

inches and outside diameter 0.2502 inches. The computed flow for the slit is 1.0004 that for the annulus. The formulae are on page 211 of de Nevers, *Fluid Mechanics for Chemical Engineers*, 2e, McGraw-Hill, 1990.

10.15 From that example we know that the required vapor pressure is less than or equal to 7.35 E-4 psia = 0.0380 mm Hg. Solving the Antoine equation for T, and inserting values we find

$$\log p = A - \frac{B}{(T+C)}; \qquad T = \frac{B}{(A - \log p)} - C$$

$$T = \frac{1343.943}{(6.95334 - \log\ 0.0380)} - 219.377°\text{C} = -59.87°\text{C} \approx -60°\text{C}$$

10.16 The required heat removal rate in the gas cooler (condenser) is

$$\dot{Q} = \dot{n}C_p\Delta T = 1000\ \text{scfm} \cdot \left(\frac{\text{lbmol / min}}{385.3\ \text{scfm}}\right) \cdot 6.95 \frac{\text{Btu}}{\text{lbmol} \cdot °\text{R}} \cdot (-74 - 100)°\text{F}$$

$$= \text{-3.14 E3} \frac{\text{Btu}}{\text{min}}$$

If we assume that the heat rejection temperature of the refrigerator is 100°F, that it must supply the refrigerant at 20°F below the exit temperature of the gas, (20°F approach temperature) and that its overall efficiency is 0.9, then its coefficient of performance (COP) will be

$$\text{COP} = \eta \frac{1}{T_h / T_l - 1} = 0.9 \frac{1}{560°\text{R} / 366°\text{R} - 1} = 1.698$$

and

$$\text{Po} = \frac{-\dot{Q}_L}{\text{COP}} = \frac{3.14\ \text{E3 Btu / min}}{1.698} \cdot \frac{60\ \text{min}}{\text{hr}} \cdot \frac{\text{HP} \cdot \text{hr}}{2545\ \text{Btu}}$$
$$= 43.5\ \text{HP} = 32.5\ \text{kw}$$

This is a fairly big refrigerator, treating a fairly small gas flow. This shows why this approach is not widely used for large gas flows!

10.17 See Prob. 10.15. For 1 ppm, p/P = 1 E-6, p = 14.7 E-6 psia = 760 E-6 mm Hg.

$$T = \frac{1343.943}{(6.95334 - \log\ 7.6\ \text{E-4})} - 219.377°\text{C} = -85.8°\text{C}$$

which would be substantially more expensive than - 60 ° C.

10.18 Using the values from that example,

$$\begin{pmatrix}\text{Permitted}\\ \text{emissions}\end{pmatrix} = 0.01295\frac{\text{mol gasoline}}{\text{mol mix}}\cdot\frac{10}{35} = 0.00370\frac{\text{mol gasoline}}{\text{mol mix}}$$

$$y_{\text{gasoline in exit gas}} = \frac{0.00370}{0.00370 + 592} = 0.00621$$

$$p = 0.00621 \cdot 14.7 \text{ psia} = 0.0913 \text{ psia}$$

$$T = \frac{5236.5}{(11.724 - \ln 0.0913)} = 370.9°\text{R} = -89°\text{F}$$

Thus the final stage refrigeration must operate 37°F colder than in the example. The rest of the example is unchanged, as is the calculation method.

10.19 (a) By straightforward elimination (on a spreadsheet)we find B = 11027.3378, and then A = 19.9896647. Then one may back calculate the two input pressures finding the input values and thus showing that the constants are correct.

(b) At 0°F = 460°R, this equation shows p = 0.0186 psia which is an adequate check of the reported value of 0.0185 psia.

(c) The estimated pressure is p = 0.0010 psia, which we are fairly safe in setting 0.00

10.20 Using Eq. 10.15 and a spreadsheet we can repeat the example for various temperatures as shown below

Temperature, deg C	30	0
Vapor pressure, psia	8.443	2.946
Mol fraction air in stream 1 = mols air	0.436	0.800
Mol fraction gasoline in stream 3	0.0318	0.0172
Vapor pressure of gasoline in stream 3, psia	0.468	0.253
Temperature of stream 3, °R	419.	400
Temperature of stream 3, °F	- 40.5	- 60.2

The answers to parts (a) and (b) are -40.5°F and -60.2°F. This is a counter-intuitive result, but is right. In winter, when the air is at 0°C, there is more air displaced per volume of gasoline displaced because there is less gasoline in the headspace, and conversely in summer. The inverse of this is that the gasoline recovered in the first separator is greater in summer than in winter, per volume of gasoline filled.

Here I have not corrected for the fact that a liter of gasoline charged displaces fewer mols of gasoline-air mix in summer than in winter, because of the higher temperature (and lower molar density) of the headspace gas. That effect would influence the above results less than does the change in vapor pressure, and in the same direction. If one repeats the calculation taking that into account, which some of your students may do, one finds

Temperature, deg C	30	0
Vapor pressure, psia	8.443	2.946
Mol fraction air in stream 1 = mols air	0.436	0.800
Mol fraction gasoline in stream 3	0.0329	0.0160

Vapor pressure of gasoline in stream 3, psia	0.484	0.235
Temperature of stream 3,°R	420.	398
Temperature of stream 3, °F	- 39.4	- 62.3

(c) The refiners raise the gasoline vapor pressure in winter and lower it in summer, which has the effect of moving the two temperatures calculated here in the direction of Example 10.10, i.e. toward a temperature of - 50°F. If they could, the refiners would deliver a gasoline which had the same vapor pressure **at engine inlet temperature** at all times. Because of the swings in ambient (and fuel tank) temperature, they cannot do that, but they come as close as they can.

10.21 The flow diagram is

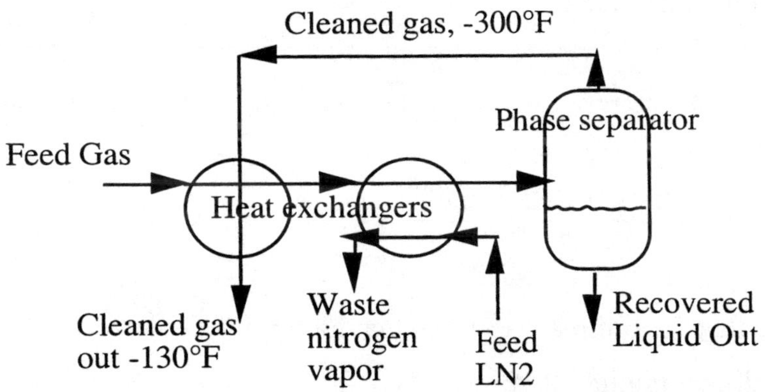

I base this on an Edwards Engineering Technical-Sales Bulletin, which shows recovering the sensible cooling of the cleaned gas, but not that of the waste nitrogen. That is an economic decision; for a large installation an additional heat exchanger to recover the sensible cooling in the gaseous nitrogen would possibly pay.

10.22 The flow diagram is

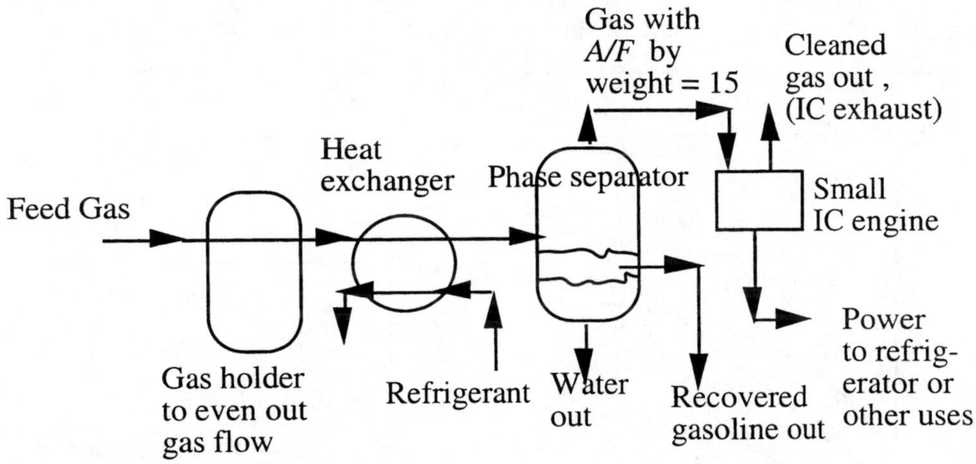

Again this is based on an Edwards Engineering Technical-Sales Bulletin.

(b) To have A/F by weight = 15, with a gasoline M = 60 g/mol, we need

$$\begin{pmatrix}\text{mol fraction}\\ \text{gasoline}\end{pmatrix} = y_{\text{gasoline}} = \frac{1/60}{1/60 + 15/29} = 0.0312$$

At one atm the corresponding vapor pressure is 14.7·0.0312 = 0.4589 psia. Solving Eq. 10.15 we find that the phase separator temperature must be - 41°F.

(c) The advantages are that instead of being a consumer of electric power to run the refrigerator(s), this plant is a net power producer. The IC engine supplies all internal power and exports power to the rest of the plant. The lowest temperature required is less than in Example 10.10, because it is not necessary to reduce the gasoline content of the vapor leaving the phase separator as much. The disadvantage is the additional complexity. The IC engine and gas holder are additional pieces of equipment, which add cost, maintenance, wiring and plumbing, etc.

10.23 See Example 10.11. (a) Here we replace the 0.010 in the log term with an 0.001, finding

$$\text{Saturation parameter} = \frac{560 \cdot 0.782}{1.8 \cdot 92} \log \frac{0.007}{0.001} = 4.88$$

From Fig. 10.11 we read that the ordinate is 32 so that $w^* = 32 \cdot \dfrac{0.782}{100} = 0.250$

(One my check this visually on Figure 10.12).

(b)

$$\text{Sat param} = \frac{760 \cdot 0.782}{1.8 \cdot 92} \log \frac{2.66}{0.005} = 9.78,$$

$$\text{ordinate} \approx 14; \quad w^* = \frac{(0.14) \cdot (0.728)}{100} = 0.109$$

Which also matches Fig. 10.11 The actual calculations leading to Figure 10.11 were done on a spreadsheet, with the values of the abscissa read as shown above.

10.24 The curves are

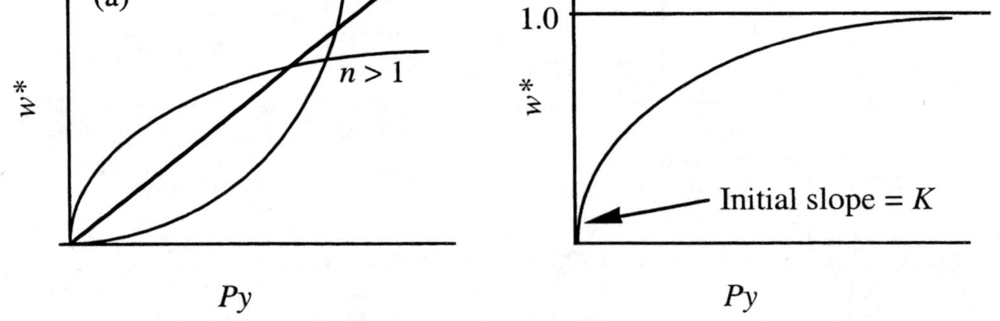

In the absorption literature curves with $n < 1$ are called "unfavorable" and those with $n > 1$ are called "favorable". Clearly a curve of the favorable type indicates that a substantial amount of adsorbate can be taken out of a very dilute gas stream, while that is not true for the "unfavorable" curve.

10.25 Here $V_s = \frac{1000\,\text{scfm}}{16\,\text{ft}^2} \cdot \frac{560°R}{528°R} = 66.3\frac{\text{ft}}{\text{min}}$; $D_p = 0.0128\,\text{ft}$

The viscosity is that of air at 100°F, = 0.022 cp. Then substituting we have

$$\frac{\Delta P}{\Delta x} = \frac{150 \cdot \left(66.3\frac{\text{ft}}{\text{min}}\right) \cdot (0.022\,\text{cp}) \cdot (0.7)^2}{(0.0128\,\text{ft})^2 \cdot (0.3)^3} \cdot 2.09\,\text{E}-5\frac{\text{lbf}\cdot\text{s}}{\text{ft}^2\cdot\text{cp}} \cdot \frac{\text{min}}{60\,\text{s}}$$

$$= 8.44\frac{\text{lbf}}{\text{ft}^2\cdot\text{ft}} = 0.059\frac{\text{psi}}{\text{ft}}; \qquad \Delta P = 0.059\frac{\text{psi}}{\text{ft}} \cdot 4\,\text{ft} = 0.23\,\text{psi}$$

Here we must check the laminar flow assumption, by calculating the packed bed Reynolds number, finding

$$\mathcal{R}_p = \frac{(0.0128\,\text{ft}) \cdot \left(66.3\frac{\text{ft}}{\text{mm}}\right) \cdot \left(0.075\frac{\text{lbm}}{\text{ft}^3} \cdot \frac{528}{500}\right)}{(0.022\,\text{cp}) \cdot (0.7)} \cdot \frac{\text{ft}\cdot\text{s}\cdot\text{cp}}{6.72\,\text{E-4 lbm}} \cdot \frac{\text{min}}{60\,\text{s}} = 96.9$$

This is too high for the Kozeny-Carman equation, but from the Ergun equation, one can see that the friction factor is roughly twice that for the Kozeny-Carman equation, so that the expected pressured drop is roughly twice this or 0.5 psi. This is probably too high for most air pollution applications, but not very high for absorption beds used in industrial situations treating gases under pressure. For an air pollution application we would probably select a wider, shallower bed, which would lead to a lower pressure drop, but also to a more rapid breakthrough. (The formulae are in section 12.1 of de Nevers, *Fluid Mechanics for Chemical Engineers*, 2e, McGraw-Hill, 1990.)

10.26 (a)

$$y_a P = 0.005\ \text{atm}; \quad y_a = \frac{0.005\ \text{atm}}{0.01\ \text{atm}} = 0.50 = 50\ \text{mol}\ \%$$

(b) $y_a P = 0.01 \cdot 0.01\ \text{atm} = 1\,\text{E-4 atm}$. From Fig. 10.12 we see that this corresponds to $w*$ 0.15 lb/lb, about half of the concentration at the start of the regeneration.

I include this problem to illustrate "pressure-swing regeneration", which is discussed at the end of Section 10.4.2.

10.27 Repeating Ex. 10.11,

$$\text{Sat param} = \frac{528 \cdot 0.782}{1.8 \cdot 92} \log \frac{0.029}{0.001} = 93.64,$$

$$\text{ordinate} \approx 38; \quad w^* = \frac{(0.38) \cdot (0.728)}{100} = 0.30$$

So at saturation we would have 0.03·100 g = 30 g = 0.33 mols adsorbed. The air flow rate is

$$Q = \frac{15 \text{ kg}}{24 \text{ hr}} \cdot \frac{\text{mol}}{29 \text{ g}} \cdot \frac{1000 \text{ g}}{\text{kg}} = 21.6 \frac{\text{mol}}{\text{hr}}$$

And the time to (ideal) breakthrough is

$$\Delta t = \frac{m}{Qy} = \frac{0.33 \text{ mol}}{\left(21.6 \frac{\text{mol}}{\text{hr}}\right) \cdot 0.001} = 15.4 \text{ hours}$$

This is an optimistic estimate, because of the assumed ideal breakthrough curve, but these masks are quite effective, as anyone who has used them knows. I have used them in a painting situation where the solvent smell was overwhelming, but the air inside the mask had no smell. When one took off the mask, the smell outside remained overwhelming.

10.28** $$x = \frac{P}{H} y = \frac{1 \text{ atm}}{40{,}100 \text{ atm}} \cdot 0.21 = 5.23 \text{ E} - 6$$

This shows that there is little dissolved oxygen in water, so that oxygen supply must be a problem for fish and other creatures that live in the water. As far as I know, all the warm-blooded creatures that live in water (seals, whales, manatees) must come to the surface to breathe air. Fish and other water dwellers are not warm blooded, which greatly reduces their oxygen requirement, which they satisfy with gills which have a large gas-exchange area.

Our blood (and that of all warm-blooded animals) has chemicals in it to make it a much better absorbent of oxygen than is pure water. In our case it is principally the hemoglobin in our red blood cells.

10.29 (a) If we choose as our basis, one mol, then

$$X = \frac{\text{mols transferable}}{1 \text{ - mols transferable}} = \frac{x}{1-x}, \text{ and also } Y = \frac{y}{1 - y}$$

(b)

$$x = \frac{0.01}{0 - 0.01} = 0.0101, \text{ an error in } x \text{ or } y \text{ of } \approx 1\%$$

For Example 10.14 the error made this way is small. This approximation would not be satisfactory for large values of x and y.

10.30** The molecular weight of n-tetradecane is given in the example, and the densities of liquid and gas at column conditions are given in Ex. 10.15.

$$\frac{Q_L}{Q_G} = \frac{L}{G} \cdot \frac{M_L}{M_G} \cdot \frac{\rho_G}{\rho_L} = 0.087 \cdot \frac{198}{29} \cdot \left(\frac{0.071 \frac{\text{lb}}{\text{ft}^3}}{47 \frac{\text{lb}}{\text{ft}^3}} \right) = 8.9\ \text{E} - 4 \approx 0.9\%$$

Yes, these values bear out those statements. The column is mostly full of gas, and its size is mostly governed by the gas flow rate.

10.31 (a)

$$x_{\text{weight}} = 5.2\ \text{E} - 4;\ x_{\text{molar}} \approx 5.2\ \text{E} - 4 \cdot \frac{18}{92} = 1.02\ \text{E} - 4$$

$$H_i = \frac{P y_i *}{x_i} = \frac{1\ \text{atm} \cdot 1}{1.02\ \text{E} - 4} = 9800\ \text{atm}$$

(b)

$$x_{\text{bottom}} = \frac{0.8 \cdot 0.005}{9800} = 4.1\,\text{E} - 7;\ \frac{L}{G} = \frac{4950\ \text{E} - 6}{4.1\,\text{E} - 7} = 1.2\ \text{E}4$$

(c) This is 14,000 times the value in the example. Clearly this would be an impossible liquid/gas flow ratio. This illustrates that one must select a solvent in which the material to be absorbed is at least fairly soluble. In general VOC are only slightly soluble in water.

10.32 The principal effect is to lower the concentration of the solvent in the cleaned gas from 50 to 15 ppm. The second effect is to increase the solvent flow rate, by the ratio of the molecular weights (212/198) = 1.07, because the mol fraction of toluene in the liquid leaving the bottom of the column will be the same as in those examples, (assuming that the Henry's law constant does not change, which is an excellent assumption!).

Examples 10.14 and 10.16 are completely unchanged. However the column diameter calculation in Example 10.15 is changed as follows. All of my calculations for those examples and for this revision of Example 10.15 are done on a spreadsheet. The physical properties are calculated from Reid, Sherwood and Prausnitz, 3e. The quantities which change are shown below

Quantity	**Examples 10-14,15,16**	**This problem**
Solvent	n-$C_{14}H_{30}$	n-$C_{15}H_{32}$
M, g/mol	198	212
ρ, g/cm^3	0.7483	0.7542
μ, cp	1.6368	1.9569
L, kg/min	20.24	21.67
L'/G'	0.1869	0.2001
β	0.00727	0.00776
α	0.230	0.231

$0.75G'_{\text{at flood}}$ lbm/(ft^2·s)	0.577	0.568
A, ft^2	2.175	2.208
D, ft	1.664	1.676

This problem illustrates the tradeoff on solvent molecular weight. We want a solvent with a low vapor pressure, but that normally means a high M. For the same mol fraction solute in the rich solution, an increased M means an increased liquid mass and volumetric flow rate, with a bigger column, bigger pumps and flow lines and bigger heat exchangers.

10.33 (a)

$$y_i{}^* = \frac{H}{P}x_i; \quad x_i = x_{\text{i bottom}} + \frac{G}{L}\left(y_i - y_{\text{bottom}}\right); \; y_i^* = \frac{H}{P}\left(x_b + \frac{G}{L}\left(y_i - y_b\right)\right)$$

$$\frac{dy_i}{\left(y_i - y_i^*\right)} = \frac{dy_i}{y_i - \frac{H}{P}\left(x_b + \frac{G}{L}\left(y_i - y_b\right)\right)} = \frac{dy_i}{y_i\left(1 - \frac{HG}{PL}\right) - \frac{H}{B}x_B + \frac{H}{P}\frac{G}{L}y_B}$$

Then the integration follows as shown to Eq. 10.27. The minus sign appears in the integration because of the minus sign in Eq. 10.17.

(b) The compact forms shown by Colburn are based on dropping some insignificant terms, see the discussion following his Eq. 11. Notice also that he runs his integrations from the top of the column (2 in his notation) to the bottom, which removes the minus sign discussed above, but also changes the sign of the ln.

(c) The plot of interest is Fig 8.20, p 310 of Treybal, *Mass Transfer Operations,* 3e, McGraw-Hill 1980, and the same plot with axes interchanged and reciprocals taken of some of the variables in *Perry's* 7e, Figure 14-5, p 14.10. To make this comparison one must solve by material balance for the concentration of toluene in the stripped solvent (at the top of the column), x_2 in those figures, x_T in the notation of this book. Also, they use m where this book uses H/P. In these examples that value is not specified. By material balance

$$x_{\text{top}} = x_{\text{bottom}} + \frac{G}{L}\cdot\left(y_{\text{top}} - y_{\text{bottom}}\right) = 0.057142 + \frac{1}{0.0868}\cdot\left(50\text{ E-6} - 5000\text{ E-6}\right)$$
$$= 1.42\text{ E-4}$$

Using this value in either of the above plots leads to N 16, as expected. The calculation is very sensitive to the value of x_{inlet}. The derivation shown in Treybal and Perry bases the y^* calculation on that inlet value, which makes it fit these plots. The method shown in this text is simpler and easier to understand, not requiring the value of the solute concentration in the stripped solvent. For industrial design one almost always chooses the other approach.

10.34 (a) See Table 7.1. LEL = 0.43·2.27% = 0.976% $y_{\text{toluene}} = \dfrac{\dot{n}_{\text{toluene}}}{\dot{n}_{\text{toluene}} + \dot{n}_{\text{air}}}$ which we rearrange to $\dot{n}_{\text{air}} = \dot{n}_{\text{toluene}} \cdot \dfrac{(1 - y_{\text{toluene}})}{y_{\text{toluene}}}$. The concentration of toluene in the final mix is small enough that $(1 - y_{\text{toluene}}) \approx 1$ so that $\dot{n}_{\text{air}} \approx \dfrac{\dot{n}_{\text{toluene}}}{y_{\text{toluene}}}$

$$\dot{n} = \frac{1000\dfrac{\text{lbm}}{\text{hr}}}{92\dfrac{\text{lbm}}{\text{lbmol}}} \cdot \frac{1}{0.00976} \cdot \frac{1}{0.1} = 1.112\text{ E4 } \frac{\text{lbmol}}{\text{hr}} = 3.2\text{ E5} \frac{\text{lbm}}{\text{hr}}$$

$$= 4.29\text{ E6 } \frac{\text{scf}}{\text{hr}} = 7.15\text{ scfm} = 33\frac{\text{m}^3}{\text{s}}$$

(b) To do this one would have to have a toluene vapor pressure in the finished product greater than that at equilibrium with toluene at the upper explosive limit. UEL = 2.27·3.22% = 7.3%. In most painting operations one is trying to get the solvent content in the finished products down to small fractions of a percent. Thus this is totally impractical. There are only a very few rare cases in which one gets out of the flammable range by going too rich. Most often one tries to be too lean.

Here you have an obvious conflict between the safety engineer and the air pollution control engineer. The safety engineer wants as much dilution as possible, to keep well below the LEL. The air pollution control engineer wants as little dilution as possible, to keep the size and cost of the control device down.

An unrelated example worth thinking about is starting up a distillation column which will fractionate liquids in the gasoline boiling range. In operation it is much too rich to burn. When it is empty and not running it is full of air. On the way up and down it passes through the combustible range. Common safety practice is to fill the column with an inert gas, either steam or nitrogen, between the running condition and the shut down condition, to avoid ever going through the combustible region.

10.35 (a)

$$k = A\exp\left(\frac{-E}{RT}\right) \qquad \frac{k_2}{k_1} = 2 = \frac{\exp\left(\dfrac{-E}{RT_2}\right)}{\exp\left(\dfrac{-E}{RT_1}\right)}$$

$$\ln 2 - \frac{-E}{RT_2} = \frac{-E}{RT_1}; \qquad \ln 2 = \frac{E}{R}\left[\frac{1}{T_1} - \frac{1}{T_2}\right]$$

$$E = \frac{R\ln 2}{\left[\dfrac{1}{T_1} - \dfrac{1}{T_2}\right]} = \frac{1.987\dfrac{\text{cal}}{\text{mol}\cdot\text{K}} \cdot \ln 2}{\dfrac{1}{293\text{K}} - \dfrac{1}{303\text{K}}} = 12.23\frac{\text{k cal}}{\text{mol}} = 51.2\frac{\text{kJ}}{\text{mol}}$$

This is typical of the reported activation energies for room temperature organic reactions.

(b) $$\ln\frac{k_2}{k_1}=\frac{E}{R}\left[\frac{1}{T_1}-\frac{1}{T_2}\right]=\frac{12{,}230\frac{\text{cal}}{\text{mol}}}{1.987\frac{\text{cal}}{\text{mol}\,K}}\left[\frac{1}{1273\text{K}}-\frac{1}{1283\text{K}}\right]=0.0376$$

$$\frac{k_2}{k_1}=\exp(0.0376)=1.038;\quad k_2\text{ is }3.8\%>k_1$$

The point of this is to get the students to look at the relationship between temperature and reaction rate. Ask them if increasing the value of *E* makes the increase of reaction rate with temperature increase or decrease? From the formula it is not obvious, but from the formula solved for *E*, it is clear. A large *E* makes a steeper reaction rate-temperature curve. So reactions involving strong bonds like the N-N bond are only practical at high temperatures. There are very few reactions which do not go faster at high temperatures; most of the ones we have studied involve some intermediate step where the equilibrium conversion decreases with increasing temperature.

10.36** Starting with Eq. 10.33 $p=\frac{c}{c_o}=\exp(-k\Delta t)$

$$\frac{\Delta t_2}{\Delta t_1}=\frac{\ln p_2}{\ln p_1}=\frac{\ln 1\,\text{E-5}}{\ln(7\,\text{E-5})}=1.58;\quad \Delta t=1.58\text{ s}$$

In incineration of hazardous materials the regulations generally call for some substance to be defined as the Principal Organic Hazardous Compound (POHC) and require that its measured destruction be some large value, normally 5 nines or 6 nines. Dioxin is present in small quantities in some Vietnam era herbicides and was suspected of being the principal cause of injury to those exposed to Agent Orange. Here its percent destruction was the specified performance measure for this incinerator.

Most engineers have heard of dioxin, but few know its structure. The structural formula is sketched at the right. It has been shown to be very potent carcinogen in animal tests.

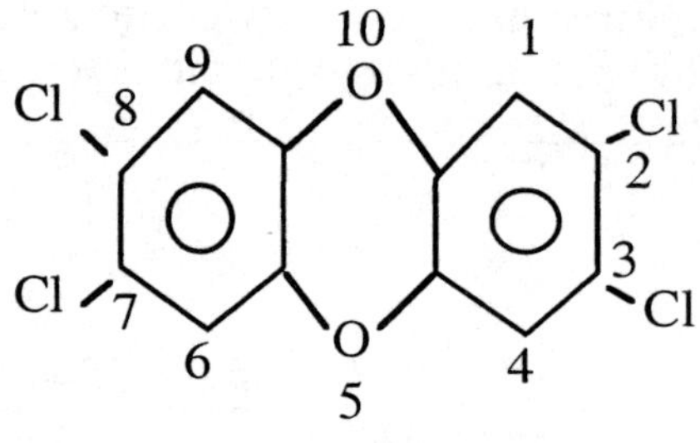

10.37 $p=\frac{c}{c_o}=\exp(-k\Delta t)$: $k=\frac{-\ln\frac{c}{c_o}}{\Delta t}=\frac{-\ln(1\,\text{E-4})}{1\text{ s}}=9.21\text{ s}$

10.38 $\Delta t=\frac{-\ln(1\,\text{E-4})}{102.27/\text{ s}}=0.08997\text{ s}\approx 0.09\text{ s}$

$$V = Q\Delta t = 1\text{ E4}\,\frac{\text{ft}^3}{\text{min}}\cdot\frac{1660°\text{R}}{528°\text{R}}\cdot 0.09\text{ s}\cdot\frac{\text{min}}{60\text{ s}} = 47.14\text{ ft}^3$$

10.39** For 99.5% destruction in 0.5 s

$$k = \frac{\ln\left(\dfrac{1}{p}\right)}{\Delta t} = \frac{\ln\left(\dfrac{1}{0.005}\right)}{0.5\text{ s}} = 10.6\,/\,\text{s};\quad k = A\exp\left(\frac{-E}{RT}\right)$$

$$T = \frac{\dfrac{-E}{R}}{\ln\left(\dfrac{k}{A}\right)} = \frac{\dfrac{-\left(56.500\dfrac{\text{cal}}{\text{mol}}\right)}{\left(1.987\dfrac{\text{cal}}{\text{mol K}}\right)}}{\ln\left(\dfrac{10.6\,/\,\text{s}}{(2.28\text{ E}13\,/\,\text{s})}\right)} = 1001\text{ K} = 1342°\text{F}$$

This is slightly higher than any of the three estimates listed in the problem, but close. Whether those other estimates are more reliable is not clear. The first-order kinetics assumed in Table 10.3 is surely a simplification of reality, but as this problem shows it does not lead to answers dramatically different from those given by more complex methods.

10.40 (a) k = 2554/s, 1400°F = 922.2 K

$$t = \frac{\ln\left(\dfrac{1}{p}\right)}{k} = \frac{\ln\left(\dfrac{1}{0.01}\right)}{25.54\,/\,\text{s}} = 0.18\text{ s}$$

(b) 5000 ppm of toluene form 35,000 ppm of CO and 20,000 ppm of H_2O. The remaining oxygen is 210,000 - (35,000 + 20,000)/2 = 182,500 ppm. If the incoming air has 20°C and 50% RH it adds 11,500 ppm of water, so that the total water is 31,500 ppm. Then taking the concentrations of water and oxygen as constant during the CO combustion (a fairly good approximation)

$$\frac{d[\text{CO}]}{[\text{CO}]} = -A\exp\left(\frac{-E}{RT}\right)\cdot[\text{O}_2]^{0.5}\cdot[\text{H}_2\text{O}]^{0.5}\,dt$$

$$\Delta t = \frac{\ln\dfrac{[\text{CO}]_i}{[\text{CO}]_f}}{A\exp\left(\dfrac{-E}{RT}\right)\cdot[\text{O}_2]^{0.5}\cdot[\text{H}_2\text{O}]^{0.5}}$$

$$A\exp\left(\frac{-E}{RT}\right) = 1.3\text{ E}14\frac{\text{cm}^3}{\text{mol}\cdot\text{s}}\cdot\exp\left(\frac{-30{,}000\dfrac{\text{cal}}{\text{mol}}}{1.987\dfrac{\text{cal}}{\text{mol}\cdot\text{K}}\cdot 298\text{ K}}\right) = 1.009\text{ E}7\,\frac{\text{cm}^3}{\text{mol}\cdot\text{s}}$$

$$\text{molar density} = 4.46\text{ E-5}\frac{\text{mol}}{\text{cm}^3}\cdot\frac{273\text{ K}}{922\text{ K}} = 1.32\text{ E-5}\frac{\text{mol}}{\text{cm}^3}$$

$$[O_2]^{0.5}\cdot[H_2O]^{0.5} = 1.32\text{E-5}\frac{\text{mol}}{\text{cm}}\cdot[0.1825]^{0.5}\cdot[0.0315]^{0.5} = 1.002\text{ E-6}\frac{\text{mol}}{\text{cm}^3}$$

$$\Delta t = \frac{\ln 100}{\left(1.009\text{ E7}\frac{\text{cm}^3}{\text{mol°s}}\right)\cdot\left(1002\text{ E-6}\frac{\text{mol}}{\text{cm}^3}\right)} = 0.46\text{ s}$$

One may estimate the significance of this difference by asking what temperature, used with the values in Table 10.3, gives a required time of 0.46 s? By simple computation one finds 1339°F. The difference between 1339 and 1400 °F is comparable to our uncertainty in all incinerator combustion calculations (see Problem 10.39).

10.41 For each component $p = \exp(-k_i\Delta t)$ thus, overall

$$p = \sum y_i \exp(-k_i\Delta t) = 0.10 = 0.25\cdot\exp\left(\frac{-1}{\text{s}}\Delta t\right) + 0.25\cdot\exp\left(\frac{-2}{\text{s}}\Delta t\right) + 0.50\cdot\exp\left(\frac{-3}{\text{s}}\Delta t\right)$$

This has no analytical solution, but is easily solved numerically, finding $t = 1.238$ s. One can then compute the individual penetrations, finding that 72% of the hydrocarbons in the effluent are A.

The instructor should comment that in real incinerators the free radicals from one compound attack another compound, so that this procedure of assuming no interaction is probably conservative, leading to a high estimate of the time and/or temperature required.

10.42 $\frac{dc}{dt} = -kc^2;\quad \frac{dc}{c^2} = -kt:\quad \left(\frac{1}{c} - \frac{1}{c_o}\right) = k\Delta t;\quad \Delta t = \frac{1}{k}\cdot\left(\frac{1}{c_f} - \frac{1}{c_o}\right)$

At 1200°F, the molar density is $4.46\text{ E-5}\frac{\text{mol}}{\text{cm}^3}\cdot\frac{492°\text{R}}{1660°\text{R}} = 1.32\text{ E-5}\frac{\text{mol}}{\text{cm}^3}$ so that

$$\Delta t = \frac{\text{mol}\cdot\text{s}}{1.8\text{ E6 cm}^3}\cdot\frac{\text{cm}^3}{1.32\text{ E-5 mol}}\left(\frac{1}{5\text{ E-6}} - \frac{1}{5\text{ E-3}}\right) = 8409\text{ s}$$

In Example 10.15, we see that for an assumed first-order reaction, using the values from Table 10.3, the calculated time for 99.9% reduction is independent of the initial concentration, and equal to 49 s for benzene at 1200°F.

For second order kinetics the time for a given percent destruction is strongly influenced by the initial concentration. If the initial concentrations had been 10 times as large (5 mol %) the calculated time for a 1000 fold reduction would be 10 times less. The time is controlled mostly by the final concentration. If one changed the initial concentration to 5 mol % in this problem, but asked for the same final concentration, the required time would be practically the same.

This time is impractically high. Fortunately one rarely has to incinerate a pure stream of a compound as difficult to incinerate as benzene. In mixtures it incinerates much more easily, because the more easily-attacked molecules supply free radicals to attack it.

10.43 The thermal temperatures are benzene > toluene > xylene; methane > propane; MEK > MIBK. All of these follow the rule that for a homologous series the required temperature decreases as the molecular weight increases. This occurs because as the molecular weight increases the amount of thermal energy stored per molecule increases. The destruction reaction presumably begins when that energy becomes high enough that if most of it is concentrated in one bond (through vibrational redistribution) it is enough to break the bond.

The catalytic temperatures do not follow as simple a pattern, probably because adsorption on the catalyst plays a major role.

These values do make chemical sense, as described above.

10.44** See example 10.19

$$\Delta t = \frac{D_o \rho}{2r} = \frac{0.001 \text{ cm} \cdot 1 \dfrac{\text{g}}{\text{cm}^3}}{2 \cdot \left(0.001 \dfrac{\text{g}}{\text{cm}^3 \cdot \text{s}}\right)} = 0.5 \text{ s}$$

10.45 This requires the students to repeat the derivation of Equation 10.34, for the cylindrical rather than spherical geometry. Let L be the length of the cylinder, which is assumed to be so much larger than the diameter that it does not change during the process. Then, writing the equivalent of the equation in Ex. 10.19 for a cylinder we have

$$-\frac{dm}{dt} = -\rho L \frac{\pi}{2} D \frac{dD}{dt} = rA = rL\pi D; \quad -\frac{dD}{dt} = 2\frac{r}{\rho}$$

$$-(D - D_o) = 2\frac{r}{\rho}(t - t_o); \quad \Delta t = \frac{\rho D_o}{2r}$$

Inserting numbers, we have

$$\Delta t = \frac{\left(1.5 \dfrac{\text{g}}{\text{cm}^3}\right) \cdot 0.001 \text{ cm}}{2 \cdot \left(0.002 \dfrac{\text{g}}{\text{cm}^3 \cdot \text{s}}\right)} = 0.0375 \text{ s}$$

The interesting result of the derivation is that the answer is the same as in Equation 10.34, the time required is $\frac{\rho D_o}{2r}$ where $\frac{D_o}{2}$ is the distance from the outside to the center. This is a natural consequence of the assumed burning rate equation, which is plausible.

10.46** See Problem 10.45 and Equation 10.34

$$-(D-D_o)=2\frac{r}{\rho}(t-t_o);\quad 1-\frac{D}{D_o}=\frac{2r\Delta t}{\rho D_o}=\frac{2\cdot\left(0.002\frac{\text{g}}{\text{cm}^3\cdot\text{s}}\right)\cdot 0.2\text{ s}}{\left(1.5\frac{\text{g}}{\text{cm}^3}\right)\cdot 0.001\text{ cm}}=0.533$$

$$\frac{D}{D_o}=1-0.533=0.467;\quad \frac{m}{m_o}=\left(\frac{D}{D_o}\right)^2=0.467^2=0.218$$

The fraction burned is 1 - 0.218 = 0.782 = 78.2%

10.47 See Problem 10.46 For $\frac{m}{m_o}=\left(\frac{D}{D_o}\right)^2=0.1,\ \left(\frac{D}{D_o}\right)=\sqrt{0.1}=0.316=1-\frac{2r\Delta t}{\rho D_o}$.

$$\Delta t=(1-0.316)\cdot\left(\frac{\rho D_o}{2r}\right)=0.683\cdot\left(\frac{\left(1.5\frac{\text{g}}{\text{cm}^3}\right)\cdot 0.001\text{ cm}^3}{2\cdot\left(0.002\frac{\text{g}}{\text{cm}^2\cdot\text{s}}\right)}\right)=0.256\text{ s}$$

10.48 See problem 10.45. Repeating that derivation for a flat plate we find the same result, viz

$$-\frac{dm}{dt}=-\rho xy\frac{dz}{dt}=rA=2rxy;\quad -\frac{dz}{dt}=\frac{2r}{\rho};\quad z_o-z=\frac{2r}{\rho}\Delta t$$

and

$$\Delta t=\frac{z_o\rho}{2r}=\frac{0.001\text{ cm}\cdot\left(2\frac{\text{g}}{\text{cm}^3}\right)}{2\cdot\left(1\text{E}-4\frac{\text{g}}{\text{cm}^2\text{s}}\right)}=10\text{ s}$$

10.49 See Problem 10.48. One may repeat the derivation, finding the same formula, with the distance being the minimum distance from surface to center. The numbers to be substituted in the equation are the same as in that problem, as is the answer.

Ask the students if the cube retains its cubical shape as it burns? It would not, but would become rounded. The corners are attacked from three directions, so that for the totally uniform burning rate assumed, they would disappear long before the burning surface reached the center.

10.50 From Equation 10.34 we see that for each particle the reduction in diameter is the same

$$D_o - D = \frac{2r\Delta t}{\rho} = \frac{2\cdot\left(0.015\,\text{E-3}\frac{\text{g}}{\text{cm}^2\cdot\text{s}}\right)\cdot(2\,\text{s})}{2\frac{\text{g}}{\text{cm}^3}} = 0.03\,\text{E-3 cm} = 0.3\,\mu$$

All particles with diameters less than 0.3 μ are totally destroyed, and all the others emerge with diameters 0.3 μ less than the diameters they entered with. We may then sketch the distribution by number, for the entering and exiting particles as shown at the right.

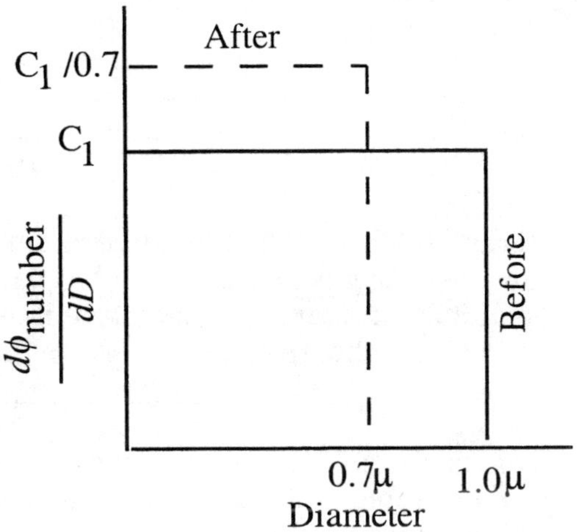

For any such particle group, we may compute the average mass of the particles from the distribution by number.

$$m_{\text{avg}} = \int_0^{D_{\max}} \rho\frac{\pi}{6}D^3\frac{d\phi_{\text{number}}}{dD}dD = \frac{\pi}{6}\rho C_1\int_0^{D_{\max}} D^3 dD = \frac{\pi}{24}\rho C_1 D^4{}_{\max}$$

For the exiting distribution the C_1 is replaced by $C_1/0.7$ as shown in the sketch, to reflect that for every 100 particles which enter only 70 exit. Then we can take the ratio of the average mass of exiting particles to entering particles, and multiply that by 0.7 to account for the reduced number of particles leaving, finding

$$\frac{m_{\text{final}}}{m_{\text{initial}}} = \frac{N_{\text{final}}}{N_{\text{initial}}}\cdot\frac{m_{\text{avg, final}}}{m_{\text{avg, initial}}} = 0.7\cdot\frac{1}{0.7}\left(\frac{D_{\text{max, final}}}{D_{\text{max, initial}}}\right)^4 = (0.7)^4 = 0.24$$

Thus 76% of the mass of the particles is burned up, while only 30% of the number of particles disappear.

(b) Using the above result, and setting the penetration to 0.05, we find

$$p = \frac{m_f}{m_i} = 0.05 = \left(\frac{D_f}{D_o}\right)^4; \quad \frac{D_f}{D_o} = (0.05)^{\frac{1}{4}} = 0.473$$

$$\Delta t = \frac{(D_o - D)\rho}{2r} = \frac{(0.572\,\text{E-4 cm})\cdot\left(2.0\frac{\text{g}}{\text{cm}^3}\right)}{2\cdot\left(0.015\,\text{E-3}\frac{\text{g}}{\text{cm}^2\cdot\text{s}}\right)} = 3.51\,\text{s}$$

This problem is only accessible to the best students in most of my classes. I put it as the last problem on an exam and the few best students in the class made reasonable starts on it. One can approach it another way using a spreadsheet. If one divides the particles into ten groups, and computes the mass-average diameter in the group entering and leaving the system numerically, then multiplies by 0.7 one finds a penetration of 0.2388, which differs but slightly from the above answer due to roundoff error.

10.51** See Example 10.19

$$r = \frac{D_o \rho}{2\Delta t} = \frac{(0.2\,\text{mm}) \cdot \left(1.2 \frac{\text{g}}{\text{cm}^3}\right)}{2 \cdot 2\,\text{s}} \cdot \frac{\text{cm}}{10\,\text{mm}} = 0.006 \frac{\text{g}}{\text{cm}^2 \cdot \text{s}}$$

This is about 400 times the rate in that example. The reasons are that coal pyrolyzes and mostly burns in the gas phase, which has a more rapid reaction rate than burning of solid carbon, and that the average temperature in the burning zone of a typical pulverized coal furnace is much higher than the 1340°F in that example.

10.52 This is a simple minimization of costs problem, complicated by the fact that we are solving for the value of the capital cost which makes the incremental cost of the catalytic unit over the thermal unit equal zero.

$$\text{Annual Incremental Costs} = (\text{C - A}) \cdot \frac{0.2}{\text{yr}} - \text{Annual Fuel Cost Saving}$$

$$\text{set} = 0, \quad (\text{C - A}) = \left(\frac{\text{Fuel saved}}{\text{yr}}\right) \cdot (\text{Fuel price}) \cdot \left(\frac{0.2}{\text{yr}}\right)$$

$$(\text{C - A}) = \frac{m C_P \Delta T}{\Delta H \text{ combustion of fuel}} \cdot \text{Fuel price} \cdot \frac{0.2}{\text{yr}}$$

$$= \frac{\left(29 \frac{\text{lb}}{\text{min}}\right) \cdot \left(0.25 \frac{\text{Btu}}{\text{lb°F}}\right) \cdot (200\text{°F})}{(359\,\text{cfm}) \cdot \left(720 \frac{\text{Btu}}{\text{ft}^3 \text{ fuel}}\right)} \cdot \frac{\$3}{1000\,\text{ft}^3 \text{ fuel}} \cdot \frac{0.2}{\text{yr}} \cdot \frac{5.25\text{ E5 min}}{\text{yr}} = \frac{\$42.50}{\text{cfm}}$$

so that we can afford to pay $42.50 per cfm more for the catalytic unit than we have to pay for the thermal afterburner.

In the case of the exhaust from automobiles, there is enough fuel present and the exhaust temperature is high enough that the catalytic unit will operate without any additional fuel.

10.53 Advantages: much lower fuel cost. Generally excellent pollutant destruction. Disadvantage: increased capital cost and complexity. These are widely used to treat streams which contain enough heating value that with their excellent heat recovery they can operate without auxiliary fuel, except for startup. They are, in effect, the same as part (b) of Figure 10.17, in which the heat content of the exit gas is transferred to the incoming gas, but in a cyclical way instead of a steady-flow way.

10.54 The relief valve is located on top of the heater, often with a tube to carry any discharged water to a floor drain. In normal operation, when the water in the heater is being heated, its thermal expansion causes some water to be expelled back into the cold water line, which is permanently open to it. If, for some reason, that line is closed, then there is no way for this thermal expansion to be accommodated, except by a rise in pressure in the water heater. This can lead to an explosive rupture of the water heater. The water line can be blocked by a pressure regulator (which some houses have to prevent excessive water pressure), by someone closing the main house inlet valve or by freezing of the water in the inlet line. The latter two are uncommon occurrences, but they occur often enough that before the relief valves were required there were occasional water heater explosions, often causing serious injuries and damage.

10.55 (a) $Q = 1000 \text{ scfm} \cdot \dfrac{560°\text{R}}{528°\text{R}} = 1061 \text{ scfm}$

$$\begin{pmatrix}\text{required volume}\\ \text{of pore space}\end{pmatrix} = Q\Delta t = 1061 \text{ scfm} \cdot 0.5 \text{ min} = 530 \text{ ft}^3$$

$$\begin{pmatrix}\text{required volume}\\ \text{of bed}\end{pmatrix} = \frac{Q\Delta t}{\varepsilon} = \frac{530 \text{ ft}^3}{0.5} = 1061 \text{ ft}^3;$$

$$\text{Area} = \frac{\text{volume}}{\text{depth}} = \frac{1061 \text{ ft}^3}{3 \text{ ft}} = 353 \text{ ft}^2; \text{ Length} = \sqrt{\text{Area}} = \sqrt{353} \text{ ft} = 19 \text{ ft on a side}$$

(b) This is very big, compared to the size of the equipment for the other ways of treating this contaminated gas stream.

10.56 Converting to the units on that plot

$$Q = 1000 \frac{\text{ft}^3}{\text{min}} \cdot \frac{60 \text{ min}}{\text{hr}} \cdot \left(\frac{\text{m}}{3.281 \text{ ft}}\right)^2 = 1800 \frac{\text{m}^3}{\text{hr}}$$

$$c = 0.005 \frac{\text{mol}}{\text{mol}} \cdot \frac{104 \dfrac{\text{g toluene}}{\text{mol}}}{29 \dfrac{\text{g air}}{\text{mol}}} = 0.0179 \frac{\text{g toluene}}{\text{g air}}$$

$$c = 0.0179 \frac{\text{g toluene}}{\text{g air}} \cdot \frac{1200 \text{ g}}{\text{m}^3} = 21.5 \frac{\text{g}}{\text{m}^3}$$

The plot suggests that we should choose scrubbing or cryocondensation

---

Chapter 11

11.1 Iron is above copper in the electromotive series, which means that if there is a shortage of oxygen, iron will take oxygen away from copper. The copper smelters operate to have the right amount of oxygen to satisfy the demands of the iron, without oxidizing the copper. The silica flux dissolves in the molten iron oxide, both lowering its melting point and its density, so that it will form a mobile liquid that floats on the copper, and can be poured off.

Equation 11.3 suggests that this is a one-stage process. In most current smelters it is a two-stage process, with most of the iron being removed in the first stage, and a liquid which is mostly CuS, called "matte", transferred to a separate vessel for the final oxidation.

The chemistry of the two reactors is different enough that they have chemically-different brick linings.

11.2 (a) At the current 300 ppm, the sulfur content of gasoline is

$$c \approx 300\ \mathrm{E}-6 \cdot 47\frac{\mathrm{lbm}}{\mathrm{ft}^3}\cdot\frac{\mathrm{ft}^3}{7.48\ \mathrm{gal}} = 0.00189\frac{\mathrm{lbm}}{\mathrm{gal}}$$

and the current emission rate (of SO_2) is

$$\begin{pmatrix}\text{Annual } SO_2 \\ \text{emission rate}\end{pmatrix} = 3.5\ \mathrm{E}8\frac{\mathrm{gal}}{\mathrm{day}}\cdot 0.00189\frac{\mathrm{lbm}}{\mathrm{gal}}\cdot\frac{64}{32}\cdot\frac{365\ \mathrm{day}}{\mathrm{yr}}\cdot\frac{\mathrm{ton}}{2000\ \mathrm{lbm}} = 2.41\ \mathrm{E}5\frac{\mathrm{ton}}{\mathrm{yr}}$$

From Table 11.2 we know that the total 1997 SO_2 emissions were 20.4 E6 tons/yr, so this is 1.2% of that total If we reduce the concentration from 300 to 40 ppm, that will decrease to 0.16 % of the total

(b) This is certainly a small fraction of the total emissions. But much of it occurs in urban areas, where the SO_2 contributes to the fine particle problem. It is also known that reducing the sulfur content will improve the performance of the catalysts in automobiles (See Chapter 13). That is one of the main reasons for this proposed regulation.

11.3 (a) The solubility (in mole fraction) is proportional to $1/H$. So the solubility of H_2S in water, expressed in mol fraction, for an equal partial pressure in the gas is (40,100/ 483) 83 times as large.
(b) First, the critical temperature of H_2S is 373.2 K, so H_2S can exist as a liquid at 20°C. As a general proposition, the lower the reduced temperature (T/T_c) the higher the solubility of the gas. In addition, H_2S reacts with water to form hydrosulfuric acid, which also increases the solubility.

As a note to the instructor: The 7th edition of Perry's shows this value of the Henry's law constant. All previous editions had an error, showing 100 times this value. If you trace it back to the International Critical Tables you will see both values at various places. The one here is almost certainly correct. I used the one from the earlier Perry's in Example 11.1 of the first edition; at the time I thought it was correct, but it wasn't.

11.4** See the next problem for the gas and liquid densities. For example 11.1

$$\frac{Q_L}{Q_G}=\frac{L}{G}\cdot\frac{M_L}{M_G}\cdot\frac{\rho_G}{\rho_L}=6.04\cdot\frac{18}{16}\cdot\frac{5\dfrac{\text{lbm}}{\text{ft}^3}}{62.3\dfrac{\text{lbm}}{\text{ft}^3}}=0.544$$

and for Example 11.2, (the above result)·(0.250/6.04) = 0.0225. These do bear out that observation; in Ex. 11.2, which is close to industrial practice, the ratio is 2%.

11.5 This is Ex. 11.3 of the first edition of this book, which is reprinted below

Example 11.3. Estimate the required column diameter in Example 11.2.

From Example 11.2, we know that $L/G = 0.250$, from which it follows that

$$\frac{L'}{G'}=\frac{LM_L/A}{GM_G/A}=\frac{L}{G}\cdot\frac{M_L}{M_G}=0.250\frac{18}{16}=0.281$$

(Here the molecular weights of pure water and pure methane are used; the true molecular weights of the absorbing solution and the natural gas differ slightly from these values.) The density of methane at 100 atm and 68°F would be 4.15 lb/ft^3 if it were a perfect gas. However, at this pressure and temperature, its density is 1.205 times the perfect gas density or 5.00 lb/ft^3. The density of pure liquid MEA is about 1.02 times that of water, so we may safely use the water density, finding

$$\beta=\frac{L'}{G'}\left(\frac{\rho_G}{\rho_L}\right)^{1/2}=0.281\left(\frac{5.00}{62.3}\right)^{1/2}=0.080$$

and $\log\beta = \log 0.080 = -1.097$ for the flooded condition. Substituting this value in Eq. (11.12), we find $\alpha = 0.15$. Then taking the liquid properties as being approximately those of water, and finding from Ref. 4 that $F \approx 50$ for typical packed columns, and using the units with the dimensions specified for Eq. (11.12), we compute

$$G'_{\text{at flooding}}=\sqrt{\frac{\alpha\rho_L\rho_G g_c}{F\Psi\mu^{0.2}}}=\sqrt{\frac{0.15\cdot 62.3\cdot 5.0\cdot 32.2}{50\cdot 1\cdot 1^{0.2}}}=5.49\frac{\text{lb}}{\text{ft}^2\cdot\text{s}}$$

and at 75 percent of flooding $G' = 4.11$ lb/(ft^2·s). From Example 11.1 we know that the gas flow rate is $\dot{m}_{\text{gas}} = 3.0$ lbmol/s, or $3\cdot 16 = 48$ lb/s. Thus

$$A=\frac{\dot{m}_{\text{gas}}}{G'}=\frac{48\text{ lb/s}}{4.11\text{ lb/ft}^2\cdot\text{s}}=11.7\text{ ft}^2=1.09\text{ m}^2$$

and

$$D=\sqrt{\frac{4}{\pi}A}=\sqrt{\frac{4}{\pi}11.7\text{ ft}^2}=3.86\text{ ft}=1.18\text{ m}$$ ■

11.6 (a) In my spreadsheet solution the columns are y, x calculated from Eq. 10.25, with the assumption that the concentrations are small enough that we can replace X and Y with x

and y, y^*, then the values at the middle of the step in y of y^*, and $(y - y^*)$, then delta h, then the running sum of the delta h's. At the top of the column the y steps are small, e.g. 4, 10, 20 .... ppm, while at the bottom they are 1000 ppm. The first step, from 4 to 10 ppm (working down the column) takes 1.60 transfer units, while the last, from 9000 to 10,000 ppm takes 0.24. The sum is 9.04 transfer units.

(b) At the top of the column $H = \frac{y^* P}{x} = \frac{0.8 \cdot 4\text{ E-}6 \cdot 100 \text{ atm}}{0.00264} = 0.121 \text{ atm}$

Here we repeat Example 10.16, using the values from this problem. The following spreadsheet shows the solution, for example 10.16, for part (b) of this problem and for part (c) of this problem For part (b) we see that $N = 9.6$ compared to the 9.04 computed in part (a). This reflects the fact that most of the column is needed for the final cleanup of the gas, so the equilibrium values at the top largely determine the column height.

Variable	**Example 10.16**	**This problem, *H* from top of column**	**This problem, *H* from bottom of column**
H atm	0.07	0.121	18.7
L/G	0.087	0.25	0.25
y top	5.00E-05	4.00E-06	4.00E-06
y bottom	5.00E-03	1.00E-02	1.00E-02
x bottom	0.057	0.0427	0.0427
P, atm	1	100	100
H/P	0.07	0.00121	0.187
HG/PL	0.8045977	0.00484	0.748
1-HG/PL	0.1954023	0.99516	0.252
numerator	4.2759E-05	7.1364E-07	-0.0005039
denominator	0.00101	0.00994833	0.0020151
fraction	0.04233527	7.1735E-05	-0.2500581
ln	-3.1621348	-9.5425369	#NUM!
N	16.1826898	9.58894743	#NUM!

(c) At the bottom $H = \frac{y^* P}{x} = \frac{0.8 \cdot 0.01 \cdot 100 \text{ atm}}{0.0427} = 18.7 \text{ atm}$ The rightmost column above shows that there is no solution with this high an H. The numerator of the fraction becomes negative.

11.7 From Ex. 11.6 we know that $N = 9.04$, and from Ex. 11.5 that $A = 11.7 \text{ ft}^2$, so that

$$h = \frac{NG}{KaPA} = \frac{9.04 \cdot 30 \frac{\text{lbmol}}{\text{s}}}{4 \frac{\text{lbmol}}{\text{ft}^3 \cdot \text{atm} \cdot \text{hr}} \cdot 100 \text{ atm} \cdot 11.7 \text{ ft}^2} \cdot \frac{3600 \text{ s}}{\text{hr}} = 20.9 \text{ ft}$$

11.8** This is an easy units problem. Basis: 1 dscm

$$\text{mols } H_2S = \frac{0.230 \text{ g}}{34 \frac{\text{g}}{\text{mol}}} = 0.00676 \text{ mols}; \quad y = \frac{0.00676 \text{ mols}}{41.57 \text{mols}} = 163 \text{ ppm}$$

This value is based on a standard temperature of 20°C (see inside the back cover). EPA standards are mostly based on a standard temperature of 25°C, which changes this from 163 to 165 ppm.

This leads to the discussion of the fact that taking hydrogen sulfide out of gases is fairly easy (if there is no other acidic component, e.g. CO_2). Of all kinds of fuels, gases are by far the easiest to desulfurize. For liquids and solids, if one wants to desulfurize, one tries to get the sulfur into a gas stream as hydrogen sulfide, then remove that. The value found is not particularly restrictive; natural gas normally has its H_2S content reduced below 4 ppm.

The typical smell of an oil refinery is due to the sulfur bearing gases it processes; it is hard to keep the leakage rates of those to zero. The petroleum engineering folklore says that when Exxon wanted to get the permits to build a new refinery at Benicia, CA in the 1960s, they promised the city fathers that one would never be able to smell the refinery. They apparently succeeded in doing that. I once complimented an Exxon engineer on showing that one could have a refinery that was a good neighbor, and make money. He told me that they had shown you could do at least one of those things. To keep their promise they apparently built two of all the major pieces of sulfur handling equipment, so that if one broke down, they had a spare ready to use. That is often done for critical pieces of equipment in refineries, e.g. reflux pumps. It is not often done for air pollution control equipment, and added significantly to the cost of that refinery.

11.9 (a) The capital cost is easy, because that plot shows straight lines on log-log paper. That plot can be represented very well by

$$\text{Capital cost } \approx \$22{,}363 \left(\dot{m}_{SO_2} \left(\frac{\text{lb}}{\text{hr}} \right) \right)^{0.65} \cdot \left(\frac{y_{SO_2}}{0.08} \right)^{-0.66}$$

If one replaces the curves for operating cost with "best-fit" straight lines, one sees that they are not parallel, as the lines in the capital cost figure are. The slopes range from 0.59 for 8% SO_2 to 0.84 for 2% SO_2. So they would have to be fit by an equation of the form

$$\text{Operating cost} \approx \text{ A}\left(\dot{m}_{SO_2}\right)^b \cdot \left(\frac{y_{SO_2}}{0.08} \right)^c$$

where b would be a function of SO_2 concentration.

(b) The total sulfuric acid produced is

$$\dot{m} = (0.8) \cdot \left(365 \frac{\text{day}}{\text{yr}} \right) \cdot \left(\frac{24\text{hr}}{\text{day}} \right) \cdot \left(\frac{10{,}000 \text{ lb } SO_2}{\text{hr}} \right) \cdot \left(\frac{98}{64} \right) \cdot \left(\frac{\text{ton}}{2000 \text{ lb}} \right) = 53{,}655 \ \frac{\text{ton } H_2SO_4}{\text{yr}}$$

For 2% SO_2 I read the capital cost as \$22 M, and the operating cost as \$4.0 M/yr. So the total annual cost would be

$$\text{Total Annual Cost} = \$4\,\text{M} + 0.2\cdot(\$22\,\text{M}) = \$8.4\,\frac{\text{M}}{\text{yr}}$$

and the cost per ton would be $\dfrac{\text{TAC}}{\dot{m}} = \dfrac{\$8.4\dfrac{\text{M}}{\text{yr}}}{0.0537\dfrac{\text{Mt}}{\text{yr}}} = \dfrac{\$156}{\text{ton}}$

For 4 and 8 % SO_2 I read the capital and operating costs as \$13 M and \$9 M, and \$1.3 M/yr and \$1 M/yr, which lead to prices per ton of \$72 and \$52. These are 1980 dollars! Currently sulfuric acid sells (in modest quantities) for about \$75/ton. Large lots probably sell for less.

11.10** The example makes clear that this is the off gas from a pyrites roaster. The reason for the sum of the oxygen and sulfur dioxide concentrations not equaling 21% is that some of the oxygen has been consumed oxidizing the iron. The concentrations shown (from the HWR example) suggest that the iron is oxidized to a mixed oxide, with more oxygen than pure FeO.

Here, if we do not take the change in number of moles into account, then as shown below the calculated conversion is 95.76%. The HWR example takes the 3.5% decrease in number of moles into account, finding 95.85% conversion. It is unlikely that we know any of the values in this problem well enough to make the difference between these numbers significant. First, let x = mols SO_2, then

$$K = 85 = \frac{x}{(0.78 - x)\cdot\left(0.108 - \dfrac{x}{2}\right)^{\frac{1}{2}}}$$; by any numerical method we find that x = 0.07469, so that $\text{Conversion} = \dfrac{0.07469}{0.0780} = 0.9576$

11.11 Mostly because as we go up in feed gas SO_2 concentration we go down in feed gas O_2 concentration. (In most sulfuric acid plants the SO_2 is produced by burning sulfur, by the reaction $S + O_2 \rightarrow SO_2$, which consumes oxygen.) We may write

$$K = \frac{y_{SO_3}}{\left(y_{SO_2}\cdot y_{O_2}^{\frac{1}{2}}\right)};\quad \text{conversion} = \frac{y_{SO_3}}{\left(y_{SO_2} + y_{SO_3}\right)} = \frac{K y_{O_2}^{\frac{1}{2}}}{\left(1 + K y_{O_2}^{\frac{1}{2}}\right)}$$

which shows that at the oxygen concentration declines, the equilibrium conversion declines as well.

11.12**(a) $p = \dfrac{0.003}{0.02} = 0.15;\quad \eta = 1 - 0.15 = 0.85 = 85\%$

(b) The answer depends on the final oxygen concentration. If we use the values from Problem 11.11, and assume 100% conversion, then the remaining oxygen (per mole of feed) will be $y_{O_2} = 0.108 - \dfrac{0.078}{2} = 0.069$

and $$K = \frac{y_{SO_3}}{y_{SO_2} \cdot y_{O_2}^{1/2}}; \quad \frac{y_{SO_3}}{y_{SO_2}} = K y_{O_2}^{1/2} = 300 \cdot (0.069)^{1/2} = 79$$

$$p_{\text{equilibrium}} = \frac{y_{SO_2}}{y_{SO_3} + y_{SO_2}} = \frac{\dfrac{y_{SO_3}}{79}}{y_{SO_3}\left(1 + \dfrac{1}{79}\right)} = 0.012; \ \eta_{\text{eqilibrium}} = 98.8\%$$

$$\frac{\eta}{\eta_{\text{equilibrium}}} = \frac{85\%}{78.8\%} = 86\%$$

The equilibrium constant in this example is from the chapter on "Contact Sulfuric Acid" in K.A. Kobe, *Inorganic Process Industries*, Macmillan, N.Y., 1948.

11.13 The capital plus operating costs of a sulfuric acid plant are proportional to 1/the SO_2 content of the feed gas. Thus going from 4% to 40% SO_2 should reduce those costs by about a factor of 10. There are other advantages and disadvantages of flash smelting; the oxygen is expensive, but the production rate is high, etc.

11.14 In general, for any adiabatic reaction $0 = \Delta H_{reaction} \cdot \Delta x + CP\Delta T; \quad \dfrac{\Delta T}{\Delta x} = \dfrac{-\Delta H}{C_P}$

Using the values in the problem we estimate $\dfrac{\Delta T}{\Delta x} = \dfrac{-\left(-23{,}000 \dfrac{\text{cal}}{\text{mol}}\right)}{7 \dfrac{\text{cal}}{\text{mol °C}}} = 3280°\text{C}$

So that for 10% SO_2 in the feed gas we would expect the slope to be 10% of this or 328°C for complete conversion. The slopes of the lines on the figure are roughly 288°C for complete conversion, which is in reasonable agreement with the value calculated above.

11.15**(a) One pound of sulfur makes (98/32) pounds of sulfuric acid, so that (sales minus raw material costs) are

$$(\text{Sales - R.M.}) = \frac{\$75}{\text{ton}} - \left(\frac{32}{98}\right)\frac{\$45}{\text{ton}} = \frac{\$60.31}{\text{ton of acid}}$$

If this is exactly equal to the capital and operating costs, you break even.

(b) Five hundred miles · \$0.03/ton mile = \$15/ton. If this is per ton of contained sulfur then there are (98/32) tons of acid and the cost is \$46/ton of contained sulfur.

(c) It costs roughly 1/3 as much to ship sulfur as to ship the equivalent amount of sulfuric acid. For this reason, if one needs a large amount of acid, one most often ships the sulfur to the place where the acid is needed, and makes the acid there.

11.16 This problem is not particularly hard for Chemical Engineers, but if the class contains many students from other disciplines, it would be well to provide some discussion of stoichiometry and the use of material balances before assigning it.

As described inside the back cover, the coal should be assumed to be the "typical Pittsburgh seam coal".

For the whole problem I choose as my basis 1 ton of dry coal fed to the boiler, and use the values in Example 7.10. I do not take into account the 5% of the SO_2 collected by the bottom ash (see Problem 4.13).

The molecular weights are S, SO_2, $CaCO_3$, $CaSO_3 \cdot 0.5\,H_2O$, $CaSO_4 \cdot 2H_2O$: 32, 64, 100, 129, 172.

(a) $$\begin{pmatrix}\text{S contained}\\ \text{in coal}\end{pmatrix} = 0.016 \cdot \frac{2000\text{ lb}}{\text{t}} = 32 \cdot \frac{\text{lb}}{\text{t}} = 1.0 \frac{\text{lbmol}}{\text{ton coal}}$$

$$\begin{pmatrix}CaCO_3\\ \text{fed}\end{pmatrix} = 1.1 \cdot \text{Stoichiometric} = 1.1 \cdot 1 \frac{\text{lbmol}}{\text{ton coal}} = 1.1 \frac{\text{lbmol}}{\text{ton coal}} = 110 \frac{\text{lb}}{\text{ton coal}}$$

(b) $$CaSO_4 \cdot 2H_2O = 1.00 \frac{\text{lbmol } SO_2}{\text{ton coal}} \cdot 0.90 \cdot 0.95 = 0.885 \frac{\text{lbmol}}{\text{ton coal}} = 147 \frac{\text{lb}}{\text{ton coal}}$$
$$CaSO_3 \cdot 0.5H_2O = 1.00 \frac{\text{lbmol } SO_2}{\text{ton coal}} \cdot 0.90 \cdot 0.05 = 0.045 \frac{\text{lbmol}}{\text{ton coal}} = 5.805 \frac{\text{lb}}{\text{ton coal}}$$
$$CaCO_3 = 1.00 \frac{\text{lbmol } SO_2}{\text{ton coal}} \cdot (1.1 - 0.9) = 0.20 \frac{\text{lbmol}}{\text{ton coal}} = 20 \frac{\text{lb}}{\text{ton coal}}$$

The total is 147 + 5.8 + 20 = 172.8 lb dry solids/ton of coal, and the weight fractions are 0.851, 0.033 and 0.116.

(c) The filter cake contains 172 lb dry solids / ton of coal. That is 90 wt % solids so that the total weight is 172.8 /0.9 = 192 lb, and thus the contained water is 19.2 lb/ton of coal.

(d) From Example 11.6 we know that we evaporate 0.065 lb water/lb of gas. From Example 7.10 we know that the total gas flow is

$$\dot{m}_{gas} = \frac{44.26\text{ lbmol}}{100\text{ lb coal}} \cdot \frac{2000\text{ lb}}{\text{t}} \cdot \frac{29\text{ lb}}{\text{lbmol}} = 25{,}671 \frac{\text{lb}}{\text{ton of coal}}$$

thus the evaporated water is

$$\dot{m}_{\text{water, evaporated}} = 25{,}671 \frac{\text{lb}}{\text{ton coal}} \cdot 0.065 \frac{\text{lb}}{\text{lb}} = 1669 \frac{\text{lb water}}{\text{ton coal}}$$

(e) The total chlorine entering the system is 0.001·2000 = 2 lb/ ton of coal. If it all leaves dissolved in the water which is in the filter cake and in the water going to the treatment plant, and its concentration is 1% by weight, then the total amount of liquid water leaving the system must be

$$\dot{m}_{\text{Cl out}} = \dot{m}_{\text{pure water out}} \cdot \frac{0.01}{0.99}$$

$$\dot{m}_{\text{pure water out}} = \dot{m}_{\text{Cl out}} \cdot \frac{0.99}{0.01} = 2\frac{\text{lb}}{\text{t}} \cdot \frac{0.99}{0.01} = 198\frac{\text{lb water}}{\text{ton coal}}$$

of that amount, 19.2 leaves in the filter cake, so that the stream to waste water treatment must be 178.8 lb/ton of coal. (No Cl exits with the water evaporated!)

Students have a hard time with part (e). Discuss with them the fact that the chlorine in the coal has to go somewhere, and that this seems the only place it can go. Discuss with them the fate of NaCl in our bodies. It comes in with whatever we eat and drink. A little goes out with sweat, most goes out in urine, some in feces. The reason we can't survive on salt water is that it has a higher concentration of salt than our urine, so that to get rid of the salt we must reject more water than we took in! Some desert creatures (e.g. kangaroo rats) have much more powerful kidneys than we; they can excrete urine with a much higher salt content than ocean water, and gladly drink it.

The early scrubbers did not have the hydroclones and bleed stream to waste water treatment. The Cl mostly went out with the filter cake (which had up to 40% water in the early scrubbers). This led to high Cl contents in the circulating water, which led to severe corrosion of stainless steel part. Through much pain and suffering we learned to control the Cl content of the circulating water, thus greatly reducing that corrosion.

11.17 $$\begin{pmatrix}\text{Coal flow}\\ \text{rate}\end{pmatrix} = 500\text{ E6 W} \cdot \frac{3413\text{ Btu}}{\text{kw}\cdot\text{h}} \cdot \frac{1}{0.35} \cdot \frac{\text{lb coal}}{13{,}600\text{ Btu}} = 3.585\text{ E5}\frac{\text{lb}}{\text{h}}$$

$$= 3.585\text{ E5}\frac{\text{lb}}{\text{h}} \cdot \frac{454\text{ g}}{\text{lb}} = 1.62\text{ E8}\frac{\text{g}}{\text{h}} = 2.71\text{ E6}\frac{\text{g}}{\text{min}}$$

Then see Ex. 7.10

$$\dot{n}_{\text{gas}} = \frac{44.14\text{ mol}}{100\text{ g coal}} \cdot 2.71\text{ E6}\frac{\text{g}}{\text{min}} = 1.200\text{ E6}\frac{\text{mol}}{\text{min}} = 2.634\text{ E3}\frac{\text{lbmol}}{\text{min}} = 1.01\text{ E6 scfm}$$

$$Q_{125°\text{F}} = 1.01\text{ E6 scfm} \cdot \frac{585°\text{R}}{528°\text{R}} = 1.128\text{ E6 acfm}$$

$$A = \frac{Q}{V} = \frac{1.128\text{ E6 acfm}}{10\text{ ft / s}} \cdot \frac{\text{min}}{60\text{ s}} = 1880\text{ ft}^2$$

$$D = \sqrt{\frac{4}{\pi}A} = \sqrt{\frac{4}{\pi}1880\text{ ft}^2} = 48.9\text{ ft} = 14.9\text{ m}$$

Ref [25] shows the diameters as 13.8 m, which is 93% of this value.

(b) $$L = \frac{100\text{ gal / min}}{1000\text{ acfm at }400°\text{F}}; \quad Q_{400°\text{F}} = 1.01\text{ E6 scfm} \cdot \frac{860°\text{R}}{528°\text{R}} = 1.645\text{ E6 acfm}$$

$$L = \frac{100\text{ gal / min}}{1000\text{ acfm at }400°\text{F}} \cdot 1.645\text{ E6 acfm} = 1.645\text{ E5}\frac{\text{gal}}{\text{min}}$$

(c) $\dot{m}_{\text{coal}} = 3.585 \text{ E5} \frac{\text{lb}}{\text{h}} = 3.14 \text{ E9 } \frac{\text{lb}}{\text{yr}}$ From Prob 11.16 we know that we expect to produce 172.8 lb dry solids per ton of coal, or

$$\dot{m}_{\text{dry solids}} = 172.8 \frac{\text{lb}}{\text{ton}} \cdot 3.14 \text{ E9 } \frac{\text{lb}}{\text{yr}} \cdot \frac{\text{ton}}{2000 \text{ lb}} = 2.71 \text{ E8} \frac{\text{lb}}{\text{yr}} = 1.36 \text{ E5} \frac{\text{ton}}{\text{yr}}$$

This is the value for a solid which is 85 wt % $CaSO_4 \cdot 2H_2O$ (gypsum). The students may also answer the weight of the gypsum contained (85% of the above answer), for which you should give full credit. As far as I know the 85 % purity gypsum product is acceptable to the wallboard industry.

11.18 From Table 11.3, the oxidation air flowrate is 1.5 mols/mol of SO_2 in the feed gas. From the previous problems the SO_2 is 1.0 lbmol/ton so the oxidation air flowrate must be 1.5 lbmol of air /ton of coal.

11.19** This is a counterflow scrubber, See Equation 9.61

$$D_{\text{D}} = 3 \text{ mm}; \ V_t = 20 \frac{\text{ft}}{\text{s}}; \quad V_g = 10 \frac{\text{ft}}{\text{s}},$$

$$N_s = \frac{\rho D_p{}^2 V}{18 \mu D_b} = \frac{\left(2000 \frac{\text{kg}}{\text{m}^3}\right) \cdot \left(10^{-6} \text{m}\right)^2 \cdot \left(\frac{20}{3.28} \frac{\text{m}}{\text{s}}\right)}{18 \cdot \left(1.8 \text{ E-5} \frac{\text{kg}}{\text{m} \cdot \text{s}}\right) \cdot (0.003 \text{m})} = 0.0125$$

$$\eta_T \approx 0$$

This indicates that wet limestone scrubbers do not collect many of the fine particles in the gas, at least by the impaction mechanism. They do collect some by the diffusion mechanism, but that fraction is small.

11.20

	Typical limestone wet scrubber	Typical venturi scrubber for particulate control
Droplet size, mm	3	0.1
$(V_g - V_p)$ ft/s	10	400
Q_L / Q_G	0.009	0.001

11.21 We will use equation 9.57 and use as our basis 1 ft^2 of cross-sectional area. The scrubber is 40 ft high and has a vertical gas velocity of 10 acf/s.

$$Q_G = 10 \frac{\text{ft}}{\text{s}} \cdot 1 \text{ft}^2 = 10 \frac{\text{ft}^3}{\text{s}}; \quad \frac{Q_L}{Q_G} = \frac{100 \text{ gal / min}}{1000 \text{ acfm}_{400°\text{F}}} \cdot \frac{860 \text{ acfm}_{400°\text{F}}}{585 \text{ acfm}_{125°\text{F}}} = 0.147$$

$$Q_L = 0.147 \cdot 10 \frac{\text{ft}^3}{\text{s}} = 1.47 \frac{\text{ft}^3}{\text{s}}$$

The volume of any drop is

$$V = \frac{\pi}{6}\left(3\text{ mm}\cdot\frac{\text{ft}}{304.8\text{ mm}}\right)^3 = 9.53\text{ E-7 ft}^3$$

so that $N_D = \dfrac{1.47\text{ ft}^3/s}{(9.53\text{ E-7 ft}^3)} = 1.54\text{ E6 } /\text{s}$

Using Eq. 9.57, we find that the number of drops present at any time is

$$N_{\text{drops}} = \frac{(1.54\text{ E6/s})\cdot(40\text{ft})}{\left(20-10\left(\frac{\text{ft}}{\text{s}}\right)\right)} = 6.17\text{ E6}$$

The volume per drop is

$$V_{\text{per drop}} = \frac{40\text{ ft}^3}{6.17\text{ E6}} = 6.48\text{ E-6 ft}^3$$

which corresponds to a cube with edge length

$$L_{edge} = (6.48\text{ E-5 ft}^3)^{\frac{1}{3}} = 0.040\text{ ft} = 0.48\text{ inches}$$

The calculated value 15 times the value for the wet-dry scrubber in Ex. 11.7, which has much smaller particles. However these drops are big enough that they generate intense local turbulence, so that the mass transfer is adequate. If they were much smaller, the upward flowing gas would carry them out of the scrubber.

11.22** Basis, 1 kw. The fan power requirement is

$$\text{Po} = \frac{Q\Delta P}{\eta} = \frac{\left(0.00156\frac{\text{m}^3}{\text{s}}\right)\cdot(20\text{ in H}_2\text{O})}{0.8}\cdot\frac{\text{Pa}}{0.0040\text{ in H}_2\text{O}} = 9.75\text{ W}$$

$$= 0.00975\text{ kW} \approx 0.01\text{ kW}$$

which is about 1% of the plant's power output.

11.23. See Prob. 10.17 and 11.22. Using the same basis, and some data from those problems

$$\frac{Q}{\text{Po}} = \frac{1.01\text{ E6 scfm}}{500\text{ MW}}\cdot\frac{860°\text{R}}{528°\text{R}} = \frac{3.29\text{ E3 acfm@400°F}}{\text{MW}} = 3.29\frac{\text{acfm}_{400°\text{F}}}{\text{kW}}$$

$$Q_w = \frac{100\text{gal}}{1000\ \text{acf}_{400°F}} \cdot \frac{3.29\ \text{acfm}_{400°F}}{\text{kW}} \cdot \frac{\text{ft}^3}{7.48\text{gal}} = 0.044 \frac{\text{ft}^3}{\text{min kW}}$$

$$\frac{\text{Pump power}}{\text{plant power}} = \left(\frac{0.044\ \text{ft}^3}{\text{min kW}}\right) \cdot \left(\frac{20\ \text{lbf}}{\text{in}^2}\right) \cdot \left(\frac{0.7426\ \text{kW} \cdot \text{min}}{33{,}000\ \text{ft lbf}}\right) \cdot \left(\frac{1}{0.8}\right) \cdot \left(\frac{144\text{in}^2}{\text{ft}^2}\right) = 0.0036$$

This is 0.36% of the plant's power output, which is significant, but only 1/3 of the amount calculated in Prob. 11.22 for the fan power to overcome the gas pressure loss in the scrubber.

11.24 It is formed by the oxidation of $SO_3^{=}$ ion. It is very desirable in the holding tank and the settler, because there we want the solid to be $CaSO_4 \cdot 2H_2O$. Forced-oxidation scrubbers are designed to have it also form in the spray tower, other types are not. In the spray tower its formation can make the scaling problem worse.

The point of this problem is simply to make them look at that figure and try to understand it. I repeat that this is a very simplified version of the true chemistry of these devices.

The dissociation and Henry's law constants are similar but not identical to those in [26]; the original article [15] does not cite its sources, nor give definitions. It appears that the Henry's law constants are the reciprocals of those normally used in Chemical Engineering.

11.25 (a) $y^* = \dfrac{xH}{P}$; if $y^* \approx 0$, then $H \approx 0$

(b) $N = -\ln\left(\dfrac{y_T}{y_B}\right) = -\ln p$

(c) $N = -\ln 0.1 = 2.303$

In Example 10.16 $p = 0.1$ but $H = 0.07$ so that $N = 16.6$. In this case the equilibrium and operating lines are straight and parallel, so that the resistance is spread out uniformly over the whole separation.

11.26 (a) $H_c = 0.679 \dfrac{\text{atm}}{\text{mol / L}} = \dfrac{Py}{c} = H_x \dfrac{c}{x}$

For dilute aqueous solutions, mol fraction of solute = x (c in mol/L)/55.5, so that

$$H_x = 0.679 \frac{\text{atm}}{\text{mol / L}} \cdot 55.5\ \text{mol / L} = 37.7\ \text{atm}$$

(b) $\begin{bmatrix}\text{SO}_2 \\ \text{dissolved}\end{bmatrix} = \dfrac{yP}{H} = \dfrac{0.001 \cdot 1\ \text{atm}}{0.679\ \text{atm / (mol / L)}} = 0.001472 \dfrac{\text{mol}}{\text{L}}$

$$[\text{H}^+] = [\text{HSO}_3^-] = \left(K_1 \cdot \begin{bmatrix}\text{SO}_2 \\ \text{dissolved}\end{bmatrix}\right)^{1/2} = (0.0156 \cdot 0.001472)^{1/2} = 0.00479$$

$$[SO_3^{-2}] = \frac{K_2 \cdot [HSO_3^-]}{[H^+]} = K_2 = 6.86\text{ E-8}$$

The H+ from the second ionization is E-4 times that from the first, so that ignoring it in the first step is a satisfactory approximation.

(c) pH = - log [0.00479] = 2.32

(d) At 68°F $H_{SO_2,c} = \dfrac{1\text{ atm} \cdot 0.001}{(0.0062 + 0.0048 + 7\text{ E-8})\dfrac{\text{mol}}{\text{L}}} = 0.16\dfrac{\text{atm}}{\text{mol / L}}$

which is the same as H_x = 8.9 atm. Repeating the calculations for 125°F, one finds 0.41 (atm/(mol/L)) = 22.7 atm.

11.27 (a) If one plots the ratio of vapor pressure/concentration vs concentration, one sees the following result. The data points are circles, the curve is a simple smooth curve fit.

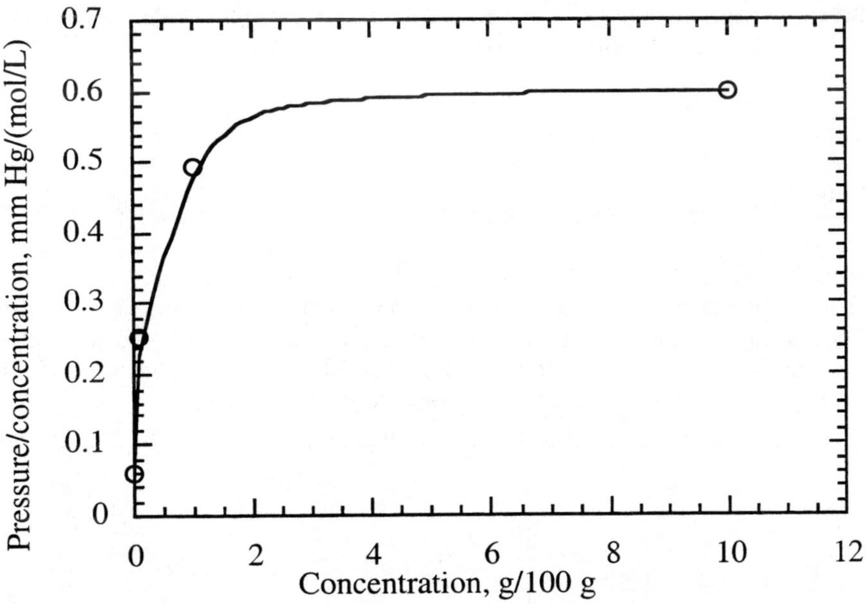

If these data points obeyed Henry's law, they would form a horizontal line on this plot, which they obviously do not. In this plot I have added an additional data point, not shown in the problem (and far out of the range of SO_2 partial pressures encountered in scrubbers), for 10 g/100 g and p = 714 torr. It adds to the picture, but the three data points in the problem clearly do not form a horizontal line.

(b) Solve Eq. 11.21 and 11.22 for [SO_2 dissolved]

$$[SO_2\text{diss}] = \frac{P y_{SO_2}}{H_{SO_2}} = \frac{[HSO_3^-] \cdot [H^+]}{K_1} \approx \frac{[HSO_3^-]^2}{K_1}$$

Then write by material balance

$$\begin{pmatrix}\text{Total SO}_2\\ \text{in all forms}\end{pmatrix} = [\text{SO}_2\text{diss}] + [\text{HSO}_3^-] = \frac{[\text{HSO}_3^-]^2}{K_1} + [\text{HSO}_3^-]$$

This is a quadratic

$$[\text{HSO}_3^-]^2 + K_1 \cdot [\text{HSO}_3^-] - K_1 \cdot \begin{pmatrix}\text{Total SO}_2\\ \text{in all forms}\end{pmatrix} = 0$$

For $K_1 = 0.0156\frac{\text{mol}}{\text{L}}$ and $\begin{pmatrix}\text{Total SO}_2\\ \text{in all forms}\end{pmatrix} = 0.00156\frac{\text{mol}}{\text{L}}$ which corresponds to 0.01 g/100 g, the positive root is 0.00143 mol/L. Subtracting that from 0.00156, we find that the dissolved SO_2 is 0.0001313 mol/L. Then

$$Py_{\text{SO}_2} = [\text{SO}_2 \text{ diss}] \cdot H_{\text{SO}_2} = 0.0001313\frac{\text{mol}}{\text{L}} \cdot 0.679\frac{\text{atm}}{\text{mol / L}}$$

$$= 8.91 \text{ E - 5 atm} = 0.068 \text{ torr}$$

Continuing the calculation on a spreadsheet we find that for the remaining concentrations the calculated pressures are 3.08 and 58.86 torr, all in excellent agreement with the reported values. For 10 g/100 g the value calculated here is 729 torr, that reported in Perry's 714.

(c) Here we want $H = \dfrac{Py}{x_{\text{all sulfur species}}}$ for the lowest concentration this is

$H = \dfrac{9.21 \text{ E - 5 atm}}{2.81 \text{ E - 5}} = 3.27$ atm and for the next two concentrations 14.16 and 27.29 atm.

For 10 g/100 g it is 33.37. The point here is to see that the simple Henry's law formulation, written in terms of the gas that went into the liquid, not taking ionization into account, give a "constant" which is far from constant. However if we take ionization into account, and base our Henry's law value on the dissolved but not ionized part of the total gas that went into the liquid, then Henry's law works very well indeed.

11.28 (a) From Ex. 11.6

$$\frac{\dot{m}_L}{\dot{m}_G} = 18.0\frac{\text{lb}}{\text{lb}}; \quad \frac{L}{G} = 18.0\frac{\text{lb}}{\text{lb}}\frac{29\frac{\text{lb}}{\text{lbmol}}}{18\frac{\text{lb}}{\text{lbmol}}} = 29\frac{\text{mol}}{\text{mol}}$$

$$\Delta y_{\text{SO}_2} = -0.009; \quad \text{thus}, \Delta x_{SO_2} = \frac{0.009}{29} = 3.10 \text{ E - 4}$$

$$\Delta c_{\text{SO}_2} = 3.10 \text{ E - 4} \cdot 55.5\frac{\text{mol}}{\text{L}} = 0.0172\frac{\text{mol}}{\text{L}}$$

(b)

$$[\text{H}^+] = 1 \text{ E - 5} + 0.0172 = 0.0172\frac{\text{mol}}{\text{L}}; \quad \text{pH} = 1.76$$

which shows that there is a significant change in pH from the top to the bottom of the scrubber.

11.29 The necessary equilibrium data are in Prob. 11.26.

$$[SO_2 \text{ diss}] = \frac{P y_{SO_2}}{H_{SO_2}} = \frac{1 \text{atm} \cdot 0.001}{1.87 \frac{\text{atm}}{\text{mol / L}}} = 5.35 \text{ E - } 4 \frac{\text{mol}}{\text{L}}$$

independent of the pH. If the pH is maintained at 4.00 by any of methods described in the problem, then, [H+] = 1 E-4,

$$[HSO_3^-] = \frac{K_1 \cdot [SO_2 \text{ diss}]}{[H^+]} = \frac{0.00678 \frac{\text{mol}}{\text{L}} \cdot 5.35 \text{ E - } 4 \frac{\text{mol}}{\text{L}}}{1 \text{ E - } 4 \frac{\text{mol}}{\text{L}}} = 0.0363 \frac{\text{mol}}{\text{L}}$$

$$[SO_3^{-2}] = \frac{K_2 \cdot [HSO_3^-]}{[H^+]} = \frac{5.35 \text{ E - } 8 \frac{\text{mol}}{\text{L}} \cdot 0.0368 \frac{\text{mol}}{\text{L}}}{1 \text{ E - } 4 \frac{\text{mol}}{\text{L}}} = 1.903 \text{ E - } 5 \frac{\text{mol}}{\text{L}}$$

and the total amount of SO_2 transferred to the liquid is 0.0363+0.00053+0.0000193 = 0.0368 mol/L.

(b) and (c) follow the same procedure, and are easy to do on the spreadsheet on which this solution was made. The results, (d) are shown in the following table, where all values except the pH are in mol/L

pH	4	5	6	2.72
H+	0.0001	0.00001	0.000001	0.00190546
SO_2 Diss	0.00053476	0.00053476	0.00053476	0.00053476
HSO_3^-	0.03625668	0.36256684	3.62566845	0.00190278
SO_3^{-2}	1.9035E-05	0.00019035	0.00190348	9.9896E-07
total S	0.03681048	0.36329195	3.62810668	0.00243854

The rightmost column is the value from Prob. 11.26. There at 20°C, pH = 2.32, while at 52°C it is 2.72, which is shown here for comparison.

Observe the 100 fold increase in total dissolved S for a pH of 6 over that for a pH of 4. This explains why the scrubber users will buy and use expensive DBA.

11.30 (a) One Ca is equivalent to 2 Na so the costs are

$$CaCO_3;\ M = 100;\ \text{cost} = \frac{\$10}{2000 \text{ lb}} \cdot \frac{100 \text{ lb}}{\text{lbmole}} = \frac{\$0.50}{\text{lb mol Ca}}$$

$$CaO;\ M = 56;\ \text{cost} = \frac{\$57}{2000 \text{ lb}} \cdot \frac{56 \text{ lb}}{\text{lbmol}} = \frac{\$1.59}{\text{lbmol Ca}}$$

$$Na_2CO_3;\ M = 106;\ \text{cost} = \frac{\$105}{2000 \text{ lb}} \cdot \frac{106 \text{ lb}}{\text{lbmol}} = \frac{\$5.56}{\text{lbmol } Na_2} = \frac{\$5.56}{\text{lb mol equiv.Ca}}$$

$$NaOH;\ M = 40;\ \text{cost} = \frac{\$370}{(2000 \cdot 0.76) \text{ lb}} \cdot \frac{40 \text{ lb}}{\text{lbmol}} = \frac{\$9.73}{\text{lbmol Na}} = \frac{\$19.47}{\text{lbmol equiv. Ca}}$$

(b) The costs are sodium hydroxide >> sodium carbonate > calcium oxide > calcium carbonate. That is the reason that once the problems with calcium carbonate were solved, most scrubbers use it.

(c) 500 miles · \$0.03/ton·mile = \$15/ ton by rail, or \$25 /ton by truck. That is significant for all but NaOH, so most scrubbers are supplied by a nearby source of limestone or lime.

11.31** From problem 11.16 we know that the limestone requirement is 110 lb/ton of coal = 0.055 ton/ ton of coal. Using the limestone price from Prob. 11.30 the cost is 0.055 ton · \$10/ton = \$0.55/ ton of coal.

See appendix C for coal prices. Compared to the cost of the coal, this is minor.

11.32 Use a steam-heated exchanger, using low- pressure steam from the boiler or inject steam directly into the plume or use a burner burning natural gas and mixing its effluent with the effluent from the scrubber. Some scrubbers are efficient enough that they can bypass some untreated hot gas around the scrubber to get plume buoyancy, and still meet their overall collection requirements. As far as I know all of these are done.

Considerable work was expended on "liquid loop reheaters" in which a heat exchanger was placed just upstream of the scrubber and another just downstream from the scrubber. Some liquid was circulated in a closed loop between these, thus using waste heat from the stack gas to reheat the plume. As far as I know the corrosion and plugging problems in these systems were too difficult for them to be economic.

11.33 For 10 μ particles of s.g. 2, the terminal settling velocity is about 0.015 ft/s. The typical velocity in the pipes coming to and leaving such devices is about 40 ft/s. So we need an average upward velocity of 0.015/40 = 0.000375, which means that the ratio of pipe diameter to scrubber diameter must be more than the square root of this or 0.019. (about a ratio of 1/50). Most are much larger than that.

11.34

(a) See sketch at the right.

(b) In counter flow the strongest absorbent (that with the lowest concentration of dissolved absorbate) contacts the cleanest gas, so that very low outlet concentrations can be obtained. In co-flow this advantage is lost.

(c) One uses co-flow when the fluid mechanics, or some other constraint prevents you from using counter flow. See the discussion on scrubbers in Section 9.2.4. Here the particles are too small to fall in counterflow, so they are carried along, making the process a co-flow process.

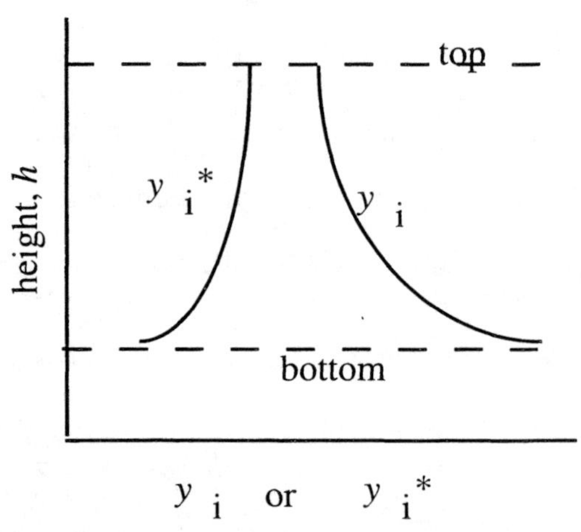

11.35 For this problem I did not take into account the 5% of the SO_2 collected by the bottom ash (see Problem 4.10).**

$$\text{Energy saved} = (100 - 20)\text{W} \cdot \frac{2000 \text{ hr}}{\text{yr}} = 1.6 \text{ E5} \frac{\text{Whr}}{\text{yr}} = 160 \frac{\text{kWh}}{\text{yr}}$$

From Ex. 7.10 100 g of coal produces 0.05 mols of SO_2. So one kWh produces

$$\frac{m}{W} = \frac{9751 \text{ Btu}}{\text{kWh}} \cdot \frac{\text{lb coal}}{13{,}600 \text{ Btu}} \cdot \frac{454 \text{ g}}{\text{lb}} \cdot \frac{0.05 \text{ mols SO}_2}{100 \text{ g coal}} \cdot \frac{64 \text{ g SO}_2}{\text{mol SO}_2} = \frac{10.4 \text{ g SO}_2}{\text{kWh}}$$

If the plant captures 90% of its SO_2 (typical of a new plant, but higher than the national average value because there are many plants in use which were built before we had serious SO_2 control requirements), then the SO_2 emissions per kWh are

$$\begin{pmatrix} \text{SO}_2 \text{ emissions} \\ \text{per unit of work} \end{pmatrix} = 10.4 \frac{\text{g SO}_2}{\text{kWh}} \cdot 0.1 = 1.04 \frac{\text{g SO}_2}{\text{kWh}}$$

and the SO_2 removal caused by this one bulb is. $\frac{160 \text{ kWh}}{\text{yr}} \cdot 1.04 \frac{\text{gSO}_2}{\text{kWh}} = 166 \frac{\text{g}}{\text{yr}} = 0.366 \frac{\text{lb}}{\text{yr}}$

This assumes that all the S in the coal goes off as SO_2. If we take into account the capture of about 5% of the S in the ash, then this answer would have to be multiplied by 0.95.

12.1**
$$E = 0.40 \text{ lb NO} \cdot \left(\frac{46 \text{ lb NO}_2}{30 \text{ lb NO}} \right) + 0.1 \text{ lb NO}_2 = 0.613 + 0.10 = 0.713 \frac{\text{lb NO}_2}{10^6 \text{Btu}}$$

which exceeds the standard. The point of this problem is for the students to see that although major sources emit both NO and NO_2, for regulatory purposes it is all assumed converted to NO_2. Regardless of the form in which the NO_X is emitted, in the atmosphere it converts to NO_2, see also Figure 1.2 and Appendix D.

12.2 (a)
$$K_p = \exp\left(-\frac{\Delta G^o}{RT} \right) = \exp\left(\frac{-2 \cdot 15.548 \frac{\text{kcal}}{\text{mol}}}{1.987 \frac{\text{cal}}{\text{mol} \cdot \text{K}} \cdot 2000\text{K}} \right) = 3.9967 \text{ E - 4} \approx 0.0004$$

(b) No effect. Equilibrium constants, *K*, are defined to be independent of pressure.

(c)
$$K_{p_1} = \exp\left(\frac{-1 \cdot 15.548 \frac{\text{kcal}}{\text{mol}}}{1.987 \frac{\text{cal}}{\text{mol} \cdot \text{K}} \cdot 2000 \text{ K}} \right) = 1.99918 \text{ E - 2} = \sqrt{K_p} \text{ in part (a)}$$

$$K_{p_{\text{part a}}} = \frac{[\text{NO}]}{[\text{N}_2]^{\frac{1}{2}} \cdot [\text{O}_2]^{\frac{1}{2}}}; \text{ NO} = K_{p_{\text{part a}}} \sqrt{[\text{N}_2] \cdot [\text{O}_2]}$$

$$= \sqrt{K_{p_{\text{part a}}} [\text{N}_2] \cdot [\text{O}_2]} = \text{Identical to Ex.12.1}$$

As expected, although the value of K_p depends on the way one chooses to write the reaction, the value of the concentrations is independent of that choice, as it must be. (The molecules do not know what we may be thinking!)

(d)
$$K_B = \exp\left(\frac{-38.002 \frac{\text{kcal}}{\text{mol}}}{1.987 \frac{\text{cal}}{\text{mol} \cdot \text{K}} \cdot 2000 \text{ K}} \right) = 7.03057 \text{ E - 5}$$

$$[\text{NO}_2] = K_B \cdot [\text{N}]^{\frac{1}{2}} \cdot [\text{O}_2] = 7.83057 \text{ E - 5} \cdot [0.78]^{\frac{1}{2}} \cdot [0.21] = 1.30394 \text{ E - 5}$$

The answers are identical, as they must be.

(e)
$$K_{12.1} = \frac{[\text{NO}_2]}{[\text{N}_2] \cdot [\text{O}_2]^{\frac{1}{2}}}; \; K_{12.3} = \frac{[\text{NO}]^2}{[\text{N}_2] \cdot [\text{O}_2]}; \; K_B = \frac{[\text{NO}_2]}{[\text{N}_2]^{\frac{1}{2}} \cdot [\text{O}_2]}$$

$$[\text{NO}] = \sqrt{K_{12.3} [\text{N}_2] \cdot [\text{O}_2]}$$

$$K_{12.1} = \frac{[NO_2]}{\sqrt{K_{12.3}[N_2]\cdot[O_2]\cdot[O_2]}} = \frac{1}{\sqrt{K_{12.3}}}\cdot\frac{[NO_2]}{[N_2]^{\frac{1}{2}}\cdot[O_2]} = \frac{K_B}{\sqrt{K_{12.3}}}$$

12.3 (a) $2x = [K_{12.3}\cdot(0.78-x)\cdot(0.21-x)]^{\frac{1}{2}}$; square both sides and regroup

$$x^2\cdot\left(\frac{4}{K_{12.3}}\right) + 0.99x - 0.1638 = 0; \quad \text{solve, } x = 3.9976\text{ E - 3},$$

$$[NO] = 2x = 7.9952\text{ E - 3}; \quad \frac{[NO]\text{ ex.12.1}}{[NO]\text{ this prob.}} = \frac{8.094\text{ E - 3}}{7.9952\text{ E - 3}} = 1.012; 1.2\%\text{ error}$$

(b) $2x = [K_{12.3}\cdot(0.78-x)\cdot(0.04-x)]$; $x = 1.72496\text{ E - 3}$; $\text{NO} = 3.449\text{ E - 3}$

$$\frac{[NO]\text{ simple}}{[NO]\text{ this prob}} = \frac{3.5327\text{ E - 3}}{3.4499\text{ E - 3}} = 1.0253; \quad 2.5\%\text{ Error}$$

We see that for the 4% oxygen case the effect is more significant. In computer programs to compute NO_X concentrations one always takes this effect into account, but for hand calculations it is often prudent to leave it out.

The JANAF tables, on which the *K* values are based claim an accuracy of ± 5%, so we should not claim an accuracy of more than ± 5% for any calculations of this type.

12.4 $[NO]_{simple} = \sqrt{K_{12.3}\cdot[N_2]\cdot[O_2]}$

$$[NO]_{complete} = 0.9\cdot[NO]_{simple} = \sqrt{K_{12.3}\cdot[N_2 - 0.5NO]\cdot[O_2 - 0.5NO]}$$

Square the second equation, divide out the $K_{12.3}$ and observe that the nitrogen concentration change is negligible, so that the [N] terms cancel and

$$0.81\cdot[O_2] = [O_2 - 0.5NO]$$

$$(1-0.81)\cdot[O_2] = [0.5NO] = 0.5\sqrt{K_{12.3}\cdot[N_2]\cdot[O_2]}$$

Square both sides and rearrange to

$$[O_2] = \frac{0.25\cdot K_{12.3}\cdot[N_2]}{0.19^2} = \frac{0.25\cdot 0.0004\cdot 0.78}{0.19^2} = 0.00216 = 0.216\%$$

This seems surprisingly small, but if one computes both the simple and complete values taking this as the starting oxygen concentration, one does find a ratio of 0.9092. If one programs the ratio on a spreadsheet and seeks the starting oxygen value which leads to a ratio of 0.90000, one finds 0.00179, which shows that the approximate solution, while close to right is not exactly right.

12.5 (a)

$$y_{N_2} = \frac{0.79 n_{dry}}{n_{total\ out}} = \frac{0.79 \cdot \left(x + \frac{y}{4} \right) \cdot \frac{(1+E)}{0.21}}{x + \frac{y}{2} + \left(x + \frac{y}{4} \right) \cdot \left[\frac{(1+E)}{0.21}(1-x) - 1 \right]}$$

Divide top and bottom by x, and let $y/x = \alpha$, simplify to

$$y_{N_2} = 0.79 \cdot \left\{ \frac{\left(1 + \frac{\alpha}{4} \right) \cdot \frac{(1+E)}{0.21}}{\frac{\alpha}{4} + \left(1 + \frac{\alpha}{4} \right) \cdot \left(\frac{1+E}{0.21} \right) \cdot (1+X)} \right\}$$

(b) If we use hydrogen as fuel $\alpha = \infty$. $1 + \frac{\alpha}{4} \to \frac{\alpha}{4}$, we factor $\frac{\alpha}{4}$ out of numerator and denominator

$$y_{N_2} = 0.79 \frac{\frac{(1+E)}{0.21}}{1 + \frac{1+E}{0.21} \cdot (1+X)}$$

$$\text{for } E = 0 \text{ and } X = 0.0116,\ y_{N_2} = 0.79 \cdot \frac{\frac{1}{0.21}}{1 + \frac{1.0116}{0.21}} = 0.647$$

(c) If we choose hydrogen as the fuel, we will have the lowest possible value of the nitrogen content of the combustion gases. For 4 % O_2, $E = 0.24$, $N_2 = 0.668$

$$\frac{\sqrt{[N_2] \cdot [O_2]}}{\sqrt{[0.78] \cdot [0.04]}} = 0.928$$

The NO_2 value is not affected because $[N_2]$ does not appear in its equilibrium relation. Again, for the most precise calculations (and for computer programs where the extra work is of no consequence) one would take this effect into account, but for hand calculations it is normally ignored.

12.6 If all of the oxygen in the atmosphere were converted to nitric acid, we would produce 63/48 = 1.31 lb of nitric acid per lb of oxygen. If that were all dissolved in the ocean, the weight fraction of nitric acid in the ocean would be

$$\frac{\text{mass HNO}_3}{\text{mass ocean}} = \frac{1}{100} \cdot 0.23 \cdot 1.31 = 0.00301 = 0.3\%$$

(The atmosphere is 21 mol % oxygen, 23 wt %!)

The equilibrium vapor pressure of NO_2 over an ocean with that much dissolved nitric acid would be

$$p_{NO_2} = 45 \text{ mmHg} \cdot (0.00301)^{7.8} = 9.76\text{ E-19 mmHg} = 1.3\text{ E-21atm}$$

From Table 12.3 we see that at 80°F if the atmosphere were at equilibrium, the concentration in the atmosphere would be 2 E-4 ppm = 2 E-10 atm. This is 10^{10} times the above amount, indicating that we would expect all the NO_2 to flow into the ocean. This caused one author (J. Ahrendts, "Reference States", *Energy, 5*, 667-677 (1980)) to seriously propose that the appropriate equilibrium reference state for the surroundings was one in which this had happened, and the atmosphere contained only traces of oxygen. Most of us have not followed his suggestion.

As discussed in [2] the apparent reason this does not occur is that there are bacteria which convert NO_2 to NH_3 which is then emitted to the atmosphere and decomposed. Without these bacteria the oxygen in the atmosphere would presumably all have gone into the oceans. This is one of many examples of creatures in our environment, of which we have known little until recently, but whose existence is of crucial importance to our survival!

The universal catalyst, if it existed would be as dangerous as the universal solvent, which could not be stored in any container. Touching the universal catalyst would turn any living thing into mostly CO_2 and water (unless the above reaction had already occurred, eliminating all living things).

12.7 (a) $[\mathrm{N}] = \sqrt{K \cdot [\mathrm{N_2}]} = \sqrt{10\ \mathrm{E}\text{-}158 \cdot (0.78)} = 8.8\ \mathrm{E}-80\ \mathrm{atm} = 8.8\ \mathrm{E}-80 \dfrac{\mathrm{mols}}{\mathrm{mol}}$

$$\text{The atmosphere is } \approx 5\ \mathrm{E18\ kg} \cdot \frac{\mathrm{mol}}{29\,\mathrm{g}} = 1.72\ \mathrm{E20\ mols, so}$$

$$\mathrm{mols\ N} = 8.83\ \mathrm{E}\text{-}80 \cdot 1.72\ \mathrm{E20} = 1.52\ \mathrm{E}\text{-}59\,\mathrm{mol}$$

$$\text{This is } 1.52\ \mathrm{E}-59\,\mathrm{mols} \cdot 6\ \mathrm{E23} \frac{\mathrm{atoms}}{\mathrm{mol}} \cong 1\ \mathrm{E}\text{-}35\ \mathrm{atoms} = 0!$$

$$\text{For } \mathrm{O_2},\ \sqrt{10\ \mathrm{E}\text{-}81 \cdot (0.21)} = 1.449\ \mathrm{E}\text{-}44\ \text{mol fraction}$$

So we would find 2.5 E - 21 mols or $\approx$ 1500 radicals in the whole atmosphere

This shows that, at equilibrium, these radicals would scarcely exist in the atmosphere. As shown in Appendix D, the O radical is formed by photochemical splitting of NO_2 and does play a significant role in the production of ozone.

(b) At 2000 K

$$[\mathrm{N}] = \sqrt{(0.78) \cdot (1\ \mathrm{E}\text{-}18)} = 8.83\ \mathrm{E}\text{-}10$$

$$[\mathrm{O}] = \sqrt{0.21 \cdot 4\ \mathrm{E}\text{-}7} = 2.9\ \mathrm{E}\text{-}4$$

$$[\mathrm{OH}] = \sqrt{0.0116 \cdot 3\ \mathrm{E}\text{-}7} = 5.9\ \mathrm{E}\text{-}5$$

In flames there are lots of these (and other) free radicals, and they are the major actors in combustion reactions.

12.8 These reaction rate expressions are from J.H. Seinfeld, "Air Pollution; Physical and Chemical Fundamentals" McGraw-Hill 1975, page 373. The *K* based on the two reaction rate can be written as

$$K = \frac{9\,\text{E}14 \cdot \exp\left(\frac{-135{,}000\,\frac{\text{cal}}{\text{mol}}}{RT}\right)}{4.1\,\text{E}13 \cdot \exp\left(\frac{-91{,}600\,\frac{\text{cal}}{\text{mol}}}{RT}\right)} = 21.9 \cdot \exp\left(\frac{-43{,}400\,\frac{\text{cal}}{\text{mol}}}{RT}\right)$$

Then with a spreadsheet one can compare its predictions to those in Table 12.3 (and ultimately on the JANAF tables) finding that over the range from 1600 to 3000 K the differences are less than 2%, but that at low temperatures, e.g. 400 K they can be as much as 20%. The reaction rate equations would not be of much use at low temperatures, so this difference is unimportant.

12.9

$$[\text{NO}]_e^2 = \frac{k_f}{k_b} \cdot [\text{N}_2] \cdot [\text{O}_2]; \quad \frac{d[\text{NO}]}{dt} = \frac{k_b \cdot [\text{NO}]_e^2}{[\text{O}_2]^{\frac{1}{2}}} - \frac{k_b \cdot [\text{NO}]^2}{[\text{O}_2]^{\frac{1}{2}}}. = \frac{k_b}{[\text{O}_2]^{\frac{1}{2}}} \cdot \left[[\text{NO}]_e^2 - [\text{NO}]^2\right]$$

Separate variables to get Eq. 12.15. This leads to integrals of the form $\int \frac{dy}{a - y^2} = b \int dx$.

There are several forms of the integral. The one we want is $\frac{1}{2a} \ln \frac{a+y}{a-y}\Big|_s^f = bx\Big|_s^f$.

Substitution the values of a and b leads to Eq 12.16.

12.10 The only hard part of the derivation is the bookkeeping in the multiplication and cancellation of terms to get Equation (F).

12.11 (a) The molar density at this condition is

$$\rho_{\text{molar}} = \left[4.46\,\text{E-5}\,\frac{\text{mol}}{\text{cm}^3}\right] \cdot \left(\frac{273\text{ K}}{2250\text{ K}}\right) \cdot \left(\frac{10\,\text{atm}}{1\,\text{atm}}\right) = 5.41\,\text{E-5}\frac{\text{mol}}{\text{cm}^3}$$

The equilibrium concentration of NO is

$$[\text{NO}]_e = [1.33\,\text{E-3} \cdot [0.78] \cdot [0.04]]^{\frac{1}{2}} = 6442\,\text{ppm}$$

In the molar density units needed here the equilibrium concentration is

$$[\text{NO}]_e = \left[5.41\,\text{E-5}\frac{\text{mol}}{\text{cm}^3} \cdot 0.006442\right] = 3.485\,\text{E-7}\frac{\text{mol}}{\text{cm}^3}$$

and

$$k_b = 4.1\,\text{E13} \cdot \exp\left(\frac{-91{,}600 \dfrac{\text{cal}}{\text{mol}}}{1.987 \dfrac{\text{cal}}{\text{mol} \cdot \text{K}} \cdot 2250\,\text{K}} \right) = 5.18\,\text{E-4} \left(\frac{\text{mol}}{\text{cm}^3} \right)^{-\frac{1}{2}} \text{s}^{-1}$$

so that

$$\alpha\, t = \frac{2 \cdot \left(3.485\,\text{E-7} \dfrac{\text{mol}}{\text{cm}^3} \right) \cdot (0.005\,\text{s}) \cdot \left(5.18\,\text{E-4} \left(\dfrac{\text{cm}^3}{\text{mol}} \right)^{1/2} \dfrac{1}{\text{s}} \right)}{\left(5.4\,\text{E-4} \cdot 0.04 \dfrac{\text{mol}}{\text{cm}^3} \right)^{\frac{1}{2}}} = 0.1226$$

$$\frac{[\text{NO}]}{[\text{NO}]_e} = \frac{1 - \exp(-0.1226)}{1 + \exp(-0.1226)} = 0.061;\ \text{NO} = 0.061 \cdot (6442\,\text{ppm}) = 394\,\text{ppm}$$

(b) This is 65 % of that value. The extended version of the Zeldovich mechanism, which takes the reactions with OH radical into account predicts higher NO values than the simple version used here, so this result is plausible.

12.12**

$$\frac{[\text{NO}]}{[\text{NO}]_e} = \frac{1500}{3615} = 0.415 = \frac{1 - \exp(-\alpha\, t)}{1 + \exp(-\alpha\, t)}$$

$$0.415 + 0.415 \exp(-\alpha\, t) = 1 - \exp(-\alpha\, t)$$

$$\exp(-\alpha t) = \frac{1 - 0.415}{1 + 0.415} = 0.4134; \quad \alpha\, t = -\ln(0.4134) = 0.883$$

$$t = \frac{0.883}{0.358/\text{s}} = 2.475\,\text{s}$$

One can also read the answer from Figure 12.4, finding the same answer (to chart reading accuracy).

12.13 From Table 12.3, $[\text{NO}]_e$= 8100 ppm. Then following Ex. 12.2, and making changes by ratios

$$\alpha = \frac{0.358}{\text{s}} \cdot \frac{\left(\dfrac{8100\,\text{ppm}}{3615\,\text{ppm}} \right)}{\sqrt{\dfrac{0.21}{0.04}}} = 0.3502 \frac{1}{\text{s}}; \quad \frac{[\text{NO}]}{[\text{NO}]_e} = \frac{1500\,\text{ppm}}{8100\,\text{ppm}} = 0.185$$

$$\exp(-\alpha\, t) = \frac{1 - 0.185}{1 + 0.185} = 0.6875; \ -\alpha\, t = \ln 0.6875 = -0.375$$

$$t = \frac{0.375}{0.350\,/\,\text{s}} = 1.07\,\text{s}$$

12.14 Reworking that example, we see that $[\text{NO}]_e$ = 8100 ppm =4.94 E-8 mol/cm^3. $[O_2]$ increases by the ratio of (0.21/0.04) to 1.281 E-6 mol/cm^3. k_b is unchanged, as is α. This latter is surprising, but $[O_2]^{0.5}$ appears in both numerator and denominator, so changing it does not change α. Thus the ratio below it does not change and [NO] = 8100 ppm· 0.173 = 1401 ppm.

12.15 This problem assumes some knowledge of heat transfer. If the class contains students who have not had a course in heat transfer, the instructor will have to provide that background if the instructor assigns the problem.

(a) $$Q = V\rho Cp\frac{dT}{dt}$$

$$= \left(1\text{cm}^3\right)\cdot\left(1.2\frac{\text{kg}}{\text{m}^3}\cdot\frac{528^\circ R}{2560^\circ R}\right)\cdot\left(\frac{6.95\,\text{cal}}{\text{mol}\cdot\text{K}}\right)\cdot\left(\frac{0.5\,\text{E}6^\circ\text{C}}{\text{s}}\right)\cdot\frac{\text{mol}}{29\,\text{g}}\cdot\frac{\text{m}^3}{1\text{E}6\,\text{cm}^3}\cdot\frac{1000\,\text{g}}{\text{kg}}$$

$$= 29.7\frac{\text{cal}}{\text{s}}$$

(b)The following calculation shows a value lower by a factor of 3. Given the approximate nature of the calculation this is fair agreement.

$$Q = Ak\frac{\Delta T}{\Delta x} = \left(3\text{ cm}^3\right)\cdot\left(0.05\frac{\text{Btu}}{\text{hr °F ft}}\right)\cdot\left(\frac{2100-68^\circ\text{F}}{0.0025\,\text{ft}}\right)\cdot\frac{\text{hr}}{3600\,\text{s}}\cdot\frac{252\,\text{cal}}{\text{Btu}}\cdot\left(\frac{\text{ft}}{30.48\,\text{cm}}\right)^2 = 9.17\frac{\text{cal}}{\text{s}}$$

(c)Assuming an emissivity of 0.1,

$$Q = A\varepsilon\sigma\left(T_1^4 - 0\right)$$

$$= \left(3\text{ cm}^2\right)\cdot(0.1)\cdot\left(\frac{0.1714\,\text{E-8 Btu}}{\text{hr °R}^4\text{ ft}^2}\right)\cdot\left(2860^\circ\text{R}\right)^4\cdot 252\frac{\text{cal}}{\text{Btu}}\cdot\frac{\text{hr}}{3600\,\text{s}}\cdot\left(\frac{\text{ft}}{30.48\,\text{cm}}\right)^2 = 2.59\frac{\text{cal}}{\text{s}}$$

which 28% of the calculated value by conduction.

(d) $$\frac{Q_c}{Q_r} = \frac{Ak\frac{\Delta T}{\Delta x}}{A\varepsilon\sigma T^4} \cong \frac{kT_1}{\varepsilon\sigma\Delta x T^4} = \frac{k}{\varepsilon\sigma\Delta x T^3};\quad \text{For } 3000^\circ F = 3460^\circ R$$

$$\frac{Q_c}{Q_r} = \frac{0.05\frac{\text{Btu}}{\text{hr °F ft}}}{(0.1)\cdot\left(\frac{0.1714\,\text{E-8 Btu}}{\text{hr ft}^2\text{ °R}^4}\right)\cdot(0.0025\,\text{ft})\cdot\left(3460^\circ\text{R}\right)^3} = 2.82$$

This shows that conduction is still the principal heat removal mechanism at this temperature, for this assumed value of the emissivity. The flames in large combustion equipment are frequently luminous, and thus have values of the emissivity greater than 0.1.

12.16 The spreadsheet program is too long to show here. The non-obvious point is that as the temperature of the gas changes, [NO] changes, even in the absence of chemical reaction, because the molar volume changes. Other than that, the procedure is straightforward.

12.17 (a)

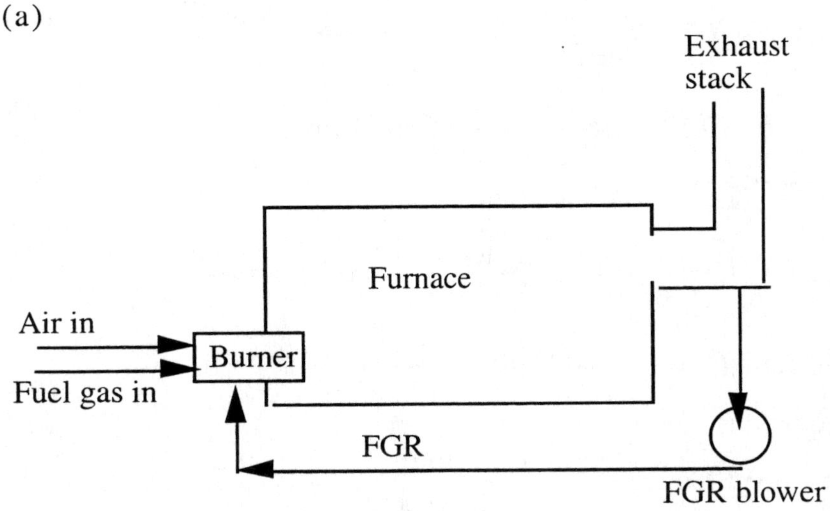

This is shown as a natural draft furnace; small furnaces mostly are. Larger ones have forced- or induced-draft fans to propel the air through the furnace. In either type an auxiliary blower is needed for the FGR, because the pressure at the base of the stack will be less than that at the burner. The kitchen stove burner shown in Fig 7.7 uses the pressure of the fuel gas to induce the air flow; the burner in Fig 12.9 apparently does not.

(b) It need not be less. The total flow of hot gas out the exhaust stack will be the same whether there is FGR or not. The lower temperature of the gases leaving the burner lead to poorer heat transfer, but the higher velocity, due to FGR leads to better heat transfer. I have seen no data on this, but I believe that there is not a substantial efficiency loss.

12.18 From Fig 12.10 we read the zero FGR values as 85, 120 and 210 ppm. From Table 7.2 we estimate that for a medium sized industrial furnace without preheating, the temperature is 3000°F. The two preheated cases would have 3232 and 3432°F. On Fig 12.5 we must interpolate the lines for 3000 and 3432°F, but after doing that, we can read the estimated concentrations as 0.5, 5.5 and 80 ppm. These are lower than the values we read from Fig 12.10 indicating that either we have estimated the temperature too low, or the time too short, or that in spite of the burner design there is some prompt NO present.

12.19 From Fig 7.4 we see that the calculated adiabatic flame temperature for methane in air 3700°F. Taking 80% of this, as discussed in the problem leads to a temperature of 2960°F. In general $\Delta T = \frac{-\Delta H_{\text{combustion}}}{m \cdot C_p}$ and 20% FGR is the equivalent of increasing m by a factor of 1.2, so the estimated temperature increase is 2960/1.2 2466°F. This is only approximate, but it shows that 20% FGR would be expected to lower the peak flame temperatures by about 500°F.

12.20** From Table 12.1, the national NO_x emission in 1997 for coal combustion was 6.242 E6 tons/yr. This is reported as if all the NO_x were in the form of NO_2 so that

$$NO_x \text{emission} = \frac{6.242 \text{ E6 tons / yr}}{46 \text{ ton / ton mol}} = 0.1357 \text{ E6} \frac{\text{ton mol}}{\text{yr}}$$

Then

$$\text{NO emission} = 0.9 \cdot 0.1357\ \text{E6} \frac{\text{ton mol}}{\text{yr}} = 0.1221\ \text{E6} \frac{\text{ton mol}}{\text{yr}}$$

$$NO_2\text{emission} = 0.1 \cdot 0.1357\ \text{E6} \frac{\text{ton mol}}{\text{yr}} = 0.0136\ \text{E6} \frac{\text{ton mol}}{\text{yr}}$$

The ammonia requirements are

$$NH_3 \text{ required for NO} = \frac{4}{6} \cdot 0.1221\ \text{E6} \frac{\text{ton mol}}{\text{yr}} = 0.0814\ \text{E6} \frac{\text{ton mol}}{\text{yr}}$$

$$NH_3 \text{ required for } NO_2 = \frac{4}{2} \cdot 0.0136\ \text{E6} \frac{\text{ton mol}}{\text{yr}} = 0.0272\ \text{E6} \frac{\text{ton mol}}{\text{yr}}$$

and the total NH_3 required is

$$NH_3 \text{ required } = 0.1086\ \text{E6} \frac{\text{ton mol}}{\text{yr}} \cdot \frac{17\text{ton}}{\text{ton mol}} = 1.84\ \text{E6} \frac{\text{ton}}{\text{yr}}$$

This is roughly 9% of the current US ammonia production.

This is a significant, but not overwhelming fraction. Most of the NH_3 produced in the world is used as a fertilizer. If we plan to use this much for NO_x control, we will need to expand our production capability. That is certainly possible, but not cheap.

12.21 See examples 7.10 and 12.5. Here we begin by working out the equivalent of Equation 12.26 for coal, finding

$$y_{O_2} = \frac{E \cdot n_{stoich}}{n_C + \frac{n_H}{2} + n_S + n_{\text{fuel}\,N_2} + n_{\text{stoich}} \cdot \left(\left(\frac{1+E}{0.21} \right) \cdot (1+x) - 1 \right)}$$

$$E = \frac{y_{O_2} \left(n_C + \frac{n_H}{2} + n_S + n_{\text{fuel}\,N_2} + n_{\text{stoich}} \cdot \left(\left(\frac{1+E}{0.21} \right) \cdot (1+x) - 1 \right) \right)}{n_{\text{stoich}} \left(1 - y_{O_2} \cdot \left(\frac{1+x}{0.21} \right) \right)}$$

Then, for the coal and air moisture content in Example 7.10 we find

$$E = \frac{0.06 \cdot \left(6.32 + 2.5 + 0.05 + 0.043 + 7.39 \cdot \left(\frac{1.0116}{0.21} - 1 \right) \right)}{7.39 \cdot \left(1 - 0.06 \cdot \left(\frac{1.0116}{0.21} \right) \right)} = 0.424 = 42.4\%$$

Then converting the emission to molar units

$$n_{\text{total out}} = 6.32 + 2.5 + 0.05 + 0.043 + 7.39 \cdot \left(\frac{1.424}{0.21} \cdot (1.0116 - 1) \right) = 52.216 \frac{\text{mol}}{100\,\text{g coal}}$$

and

$$n_{NO_2} = \frac{18\text{ lb}}{2000\text{ lb}} \cdot \frac{\text{lb mol}}{46\text{ lb}} \cdot \frac{454\text{ mol}}{\text{lb mol}} \cdot \frac{\text{lb coal}}{454\text{ g coal}} = 1.956\text{ E-4} \frac{\text{mol}}{\text{g coal}} = 0.01956 \frac{\text{mol}}{100\text{ g coal}}$$

$$y_{NO_2} = \frac{0.01956 \frac{\text{mol}}{100\text{ g}}}{52.216 \frac{\text{mol}}{100\text{ g}}} = 3.75\text{ E-4} = 375\text{ ppm}$$

Unfortunately one often sees the oxygen content of the gas stated on a dry basis, which would be found if all the water in the flue gas were condensed. This corresponds to what one found in an Orsat Analyzer, which was the state of the art in flue gas analysis in 1940.

12.22 One multiplies through by the denominator of the term on the right, groups the terms in *E*, and divides out its multiplier to find the result shown.

12.23** (a) Use Equation 12.26

$$E = \frac{0.04 \cdot \left\{1 + \frac{4}{2} + 2 \cdot \left(\frac{1.0116}{0.21} - 1\right)\right\}}{2 \cdot \left\{1 - 0.04 \cdot \left(\frac{1.0116}{0.21}\right)\right\}} = 0.263$$

$$n_{\text{total out}} = 1 + \frac{4}{2} + 2 \cdot \left[\frac{1.263}{0.21} \cdot (1.0116) - 1\right] = 13.17 \frac{\text{mol}}{\text{mol } CH_4}$$

$$c = \frac{0.000821}{13.17} = 62.5\text{ ppm}$$

(b) The mols of water are

$$n_{\text{water}} = \frac{4}{2} + \frac{2 \cdot (0.0116)}{0.21} = 2 + 0.110 = 2.110$$

so that on a dry basis the mols out are

$$n_{\text{total out, dry}} = 14.95 - 2.11 = 12.84$$

and

$$c = \frac{0.000821}{12.84} = 63.9\text{ ppm}$$

12.24 Starting with Example 12.5

$$EF = \frac{0.110 \text{ lb}}{1 \text{ E6 Btu}}; \quad \text{For methane } \Delta H_c = 21{,}502 \frac{\text{Btu}}{\text{lb}}, \text{ so}$$

$$EF = \frac{0.110 \text{ lb}}{1 \text{ E6 Btu}} \cdot \frac{21{,}502 \text{ Btu}}{\text{lb fuel}} = 0.00236 \frac{\text{lb NO}_X}{\text{lb fuel}} = 2.36 \frac{\text{g NO}_X}{\text{kg fuel}}$$

12.25** From Figure 12.4 it appears that the NO_X expressed as NO_2 is about 140 g/GJ. If all the N in the coal were converted to NO_2 the emission factor would be

$$EF = \frac{0.015 \frac{\text{lb}}{\text{lb}} \cdot \frac{46}{14}}{13{,}006 \frac{\text{Btu}}{\text{lb}}} \cdot \frac{454 \text{ g}}{\text{lb}} \cdot \frac{\text{Btu}}{1.055 \text{ kJ}} \cdot \frac{10^6 \text{ k}}{\text{G}} = 1631 \frac{\text{g}}{\text{GJ}}$$

Thus, the emission factor reported in Figure 12.4 is $\quad \text{Fraction} = \dfrac{140 \frac{\text{g}}{\text{GJ}}}{1631 \frac{\text{g}}{\text{GJ}}} = 8.6\%$

of the factor we would have for all the N being converted. This is lower than is normally reported, and suggests that perhaps the coal burned in Finland is low in N, compared to the Pittsburgh Seam Coal used in examples in this book.

12.26 From Ex. 12.5 $\quad NO_x \text{ expressed as } NO_2 = 61 + 89 \cdot \frac{46}{30} = 197 \frac{\mu\text{g}}{\text{kcal}} = 47.2 \frac{\text{g}}{\text{GJ}}$

From Table 7.2 the flame temperature in a kitchen stove is estimated at 2400°F = 1589°C. From Fig. 12.4 this corresponds to a prompt emission factor of

$35 \frac{\text{ng}}{\text{J}} = 146 \frac{\text{ng}}{\text{cal}} = 146 \frac{\mu\text{g}}{\text{kcal}} = 146 \frac{\text{g}}{\text{GJ}}$ 75% of the value in Es. 12.5 Given the various uncertainties in all these values, this is pretty good agreement.

12.27 From that problem the annual energy saving is 1.56 E6 Btu. From Table 3.1 the NSPS is 0.6 lb/ 1 E6 Btu, so that

$$\begin{pmatrix}\text{Reduction in} \\ \text{emissions}\end{pmatrix} = 1.56 \text{ E6 } \frac{\text{Btu}}{\text{yr}} \cdot \frac{0.6 \text{ lb}}{1 \text{ E6 Btu}} = 0.94 \frac{\text{lb NO}_x}{\text{yr}}$$

Chapter 13

13.1 $$\frac{\text{mower HC emissions}}{\text{auto HC emissions}} = \frac{59\frac{\text{g}}{\text{hr}}}{0.25\frac{\text{g}}{\text{mi}}\cdot 25\frac{\text{mi}}{\text{hr}}} = 9.44$$

This overstates because it compares an in-service mower to a new auto; and it does not include fueling and fuel storage emissions. It does show that small engines, particularly 2-cycle engines are large HC emitters. Currently EPA is preparing emissions regulations for small gasoline engines which will make it hard for 2-cycle engines to survive.

13.2 See table 13.2

Year	CO g/mi	NO_X g/mi	HC, tailpipe g/mi	HC, crankcase, g/mi	HC, evaporative, g/mi
1960	87	3.6	8.8	3	4
1993	3.4	0.4	0.25	0	0.27
(1960-1993)	83.6	3.2	8.55	3	3.73

$$\text{For 10,000 mi/yr, } -\Delta\text{CO} = 83.6\frac{\text{g}}{\text{mi}}\cdot 10{,}000\frac{\text{mi}}{\text{yr}} = 836{,}000\frac{\text{g}}{\text{yr}} = 836\frac{\text{kg}}{\text{yr}} = 1843\frac{\text{lb}}{\text{yr}}$$

The combined HC reduction is 8.55+3+3.73 = 15.28 g/mi. The corresponding reductions for HC and NO are 336.9 lb/yr and 70.5 lb/yr.

These values are large enough that some oil refiners in Los Angeles have bought and junked old cars, to get the HC emission reduction credits they needed to meet their permit requirements.

13.3 $$\frac{A}{F} = \frac{\left(x+\frac{y}{4}\right)\cdot(32+3.76\cdot 28)}{12x+y} = \frac{\left(1+\frac{1}{4}\frac{y}{x}\right)\cdot(137.28)}{12+\frac{y}{x}}$$

The possible range for conventional fuels is not very broad. For benzene (*y/x* =1) it is 13.2, for butane (*y/x* =2.5) it is 15.4. However for an unconventional fuel like methanol it is much less, 6.4.

13.4** (a) MTBE has M = 88, so its weight fraction of oxygen is (16/88) = 0.1818. If the gasoline is to have 2.7% oxygen then

$$\text{mass gasoline}\cdot 0.027 = \text{mass MTBE}\cdot 0.1818; \quad \frac{m_{\text{MTBE}}}{m_{\text{gasoline}}} = \frac{0.027}{0.1818} = 0.1485 \approx 15\%$$

(b)Basis 1 kg of fuel, = 0.8515 kg ordinary gasoline, 0.1485 kg MTBE

$$n_{\text{gasoline}} = \frac{851.5}{113} = 7.535; \quad n_{\text{MTBE}} = \frac{148.5}{88} = 1.687$$

$$\text{mols C} = 7.535 \cdot 8 + 1.687 \cdot 5 = 68.71; \quad \text{mols H} = 7.535 \cdot 17 + 1.687 \cdot 12 = 148.33$$

$$\text{mols O} = 1.687; \quad O_2 \text{ required} = 68.71 + \frac{148.33}{4} - \frac{1.687}{2} = 104.95 \text{ mols}$$

$$\text{Air required} = 104.95 \text{ mols } (32 + 3.76 \cdot 28) = 14.407 \frac{\text{kg}}{\text{kg}}$$

compared to 14.88 for a "typical;" gasoline.

Mostly MTBE and other oxygenated fuels work by making the engines run leaner. In a car with a mechanical fuel system the air fuel ratio is set by mechanical devices which control the volumetric ratio of liquid fuel to air. If we replace ordinary fuel which requires A/F = 14.88 with oxygenated fuel which requires A/F = 14.4, without changing the volumetric ratio of liquid to air, then the normalized A/F ratio, λ , will be increased, moving us to the right on Figure 13.2. This reduces the CO and HC emissions, and in most cases increases the NO_x emissions.

Most modern autos with three-way catalysts have feedback control of the oxygen content of the exhaust gas so that they will respond to a change in fuel oxygen content (from zero to 2.7 wt %) by increasing the fuel flow to keep the normalized A/F ratio , λ , very close to 1.00. However those cars do not operate on feedback control during cold starts nor during full-throttle acceleration, which are the times of greatest CO emissions, so that even for them oxygenated fuels provide some CO reduction.

To learn more than you want to know about the effects of oxygenated fuels, or of the other changes in gasoline which are coming about currently in Los Angeles and soon thereafter in the rest of the USA, see V.R. Burns, M.C. Ingham and H.R. Doherty, editors, "Auto/Oil Air Quality Improvement Research Program", *SP-920*, Society of Automotive Engineers, Inc. 1992.

The oil companies and EPA have an ongoing fight over proposed EPA regulations which would require the oil companies to use agriculture-based ethanol as the oxygen source in some fraction of their gasoline. This is a simple payoff to the farm states for votes on other matters in the US Senate, with no technical merit; there are lots of farm state senators. It appears that the farm states will win, but that is not certain.

The matter is further complicated by California rules requiring the phase-out of MTBE. They believe that uncombusted MTBE is getting into surface waters and causing ecological damage.

13.5 (a) Repeating the previous problem, again taking 1 kg as the basis, $M_{\text{ethanol}} = 46$ g/mol, 1 kg = 1000/46 =21.74 mols

$$n_C = 21.74 \cdot 2 = 43.48; \quad n_H = 21.74 \cdot 6 = 130.44; \quad n_O = 21.74$$

$$n_{O_2 \text{ required}} = 43.48 + \frac{130.44}{4} - \frac{21.74}{2} = 65.22 \text{ mols}$$

$$\text{mass air} = 65.22 \cdot (32 + 3.76 \cdot 28) = 8953 \text{ g} = 8.95 \text{ kg}$$

$$\frac{\text{A}}{\text{F}} = 8.95$$

For many years Brazil, which has little natural petroleum, has had a complete system of pure ethanol fueled vehicles, complete with the supply and distribution system to support them. However 16/46 = 35% by weight of ethanol is oxygen, so while the thermal efficiency of ethanol-fueled engines is satisfactory, the miles per gallon or miles per pound of fuel is much less than for gasoline.

13.6 (a) From Fig 13.2 at $\lambda = 0.95$, HC 1000 ppm.

(b) $$C_xH_y + \lambda\left(x + \frac{y}{4}\right)O_2 \rightarrow zCO + (x - z - \alpha x)CO_2 + \frac{(y - \alpha y)}{2}H_2O + \alpha C_xH_y$$

(c) $$2\lambda\left(x + \frac{y}{4}\right) = z + 2(x - z - \alpha x) + \frac{(y - \alpha y)}{2} = -z + 2x(1 - \alpha) + \frac{y}{2}(1 - \alpha)$$

(d)

$$\alpha = 1 - \frac{2\lambda \cdot \left(x + \frac{y}{4}\right) - z}{\left(2x + \frac{y}{2}\right)} = 1 - \frac{2 \cdot 0.95 \cdot \left(8 + \frac{17}{4}\right) - \frac{1.225}{2}}{\left(2 \cdot 8 + \frac{17}{2}\right)} = 0.075$$

$$n_T = x(1 - \alpha) + \frac{y}{2}(1 - \alpha) + \alpha + \lambda \cdot \left(x + \frac{y}{4}\right) \cdot \frac{0.79}{0.21}$$

$$= 8(1 - 0.075) + \frac{17}{2}(1 - 0.075) + 0.075 + 0.95 \cdot \left(8 + \frac{17}{2}\right) \cdot \frac{0.79}{0.21}$$

$$= 59.12 \text{ mols}; \quad y_{HC} = \frac{\alpha}{n_T} = \frac{0.075 \text{ mols}}{59.12 \text{ mols}} = 0.001268 = 1268 \text{ ppm}$$

(e) Within chart reading accuracy these are the same.

13.7

$$\begin{pmatrix}\text{volume}\\ \text{per}\\ \text{combustion}\end{pmatrix} = \frac{\text{gal}}{25 \text{ mi}} \cdot \frac{60 \text{ mi}}{\text{hr}} \cdot \frac{\text{hr}}{60 \text{ min}} \cdot \frac{\text{min}}{2000 \text{ rev}} \cdot \frac{2 \text{ rev}}{\text{combustion}} \cdot \frac{1}{4}\begin{pmatrix}\text{to each}\\ \text{cylinder}\end{pmatrix} \cdot \frac{3.785 \text{ L}}{\text{gal}}$$

$$= 0.03785 \text{ (cc per combustion)}$$

An ordinary chemistry laboratory buret produces about 20 drops per g, or an average volume of 0.05 cc/drop. Thus the volume per cylinder per combustion is about three-fourths of the volume of a typical drop from a typical laboratory buret.

13.8** $$\Delta T_{\text{air}} = \frac{-\lambda}{\left(mC_{\text{P, air}}\right)} \approx \frac{-140\dfrac{\text{Btu}}{\text{lb}}}{14\ \text{lb}\cdot 0.25\dfrac{\text{Btu}}{\text{lb}^\circ\text{F}}} = -\,40^\circ\text{F}$$

This decrease in temperature can cause carburetor icing. If the air is cold and wet, (e.g. foggy), this can produce ice which plugs the small holes in carburetors. This is a frequent cause of crashes when light planes take off in cold, wet conditions. Many have electric carburetor heaters to prevent this.

13.9 For combustion at constant volume $T_2 = T_1 + \dfrac{\Delta h}{mC_V}$

For compression $$\left(\frac{T_2}{T_1}\right) = \left(\frac{P_2}{P_1}\right)^{\frac{R}{C_P}}$$

For heat first, compress second

$$T_3 = \left(T_1 + \frac{\Delta h}{mC_V}\right)\left(\frac{P_2}{P_1}\right)^{\frac{R}{C_P}} = T_1\left(\frac{P_2}{P_1}\right)^{\frac{R}{C_P}} + \left(\frac{P_2}{P_1}\right)^{\frac{R}{C_P}} \frac{\Delta h}{mC_V}$$

For compress first, heat second

$$T_3 = T_1\left(\frac{P_2}{P_1}\right)^{\frac{R}{C_P}} + \frac{\Delta h}{mC_V}$$

$\left(\dfrac{P_2}{P_1}\right)^{\frac{R}{C_P}}$ is always greater than 1, so heating first and compressing second always gives a higher final temperature than the reverse. The heat capacities are not constant, which modifies this argument, but not enough to invalidate it.

To check plausibility, we write

$$T_{3,\ \text{heat-then-compress}} - T_{3,\ \text{compress-then-heat}} = \frac{\Delta h}{mC_P}\left[\left(\frac{P_2}{P_1}\right)^{\frac{R}{C_P}} - 1\right]$$

Here $\dfrac{\Delta h}{mC_P}$ 2016 K. From Prob. 13.23 we know that P_{max} 827 psia = 56.25 atm.

Before combustion the pressure should be $7^{1.4}$ = 15.2 atm, so that

$$T_{3,\ \text{heat-then-compress}} - T_{3,\ \text{compress-then-heat}} = 2016\ \text{K}\left[\left(\frac{56.25}{15.2}\right)^{\frac{1}{3.5}} - 1\right] = 911\ \text{K}$$

This is the highest plausible estimate; based on it the observation of 400 K seems quite reasonable.

13.10 The procedure is described in detail, with examples on page 401 et. seq. of D.M. Himmelblau *Basic Principles and Calculations in Chemical Engineering* Fourth Edition, Prentice-Hall 1982.

If the heat capacities of the air-fuel mix and of the combustion products were the same, then the heat of combustion would be completely independent of the temperature of combustion. However (See Taylor Ref 9, Volume 2, page 120), the product of number of mols times heat capacity per lbmol of fuel is 135.6 Btu/lbmol °R for fuel and oxygen and 153.96 But/lbmol °R for combustion products. (The nitrogen is left out of the calculation because its heat capacity is the same for heating or cooling), The difference between the two is 18.4 Btu/lbmol °R, which multiplied by (694 -77)°R = -1.13 $\cdot 10^4$ Btu/lb mole of fuel or about -100 Btu/lb of fuel which is about 0.5% of the fuel heating value. This is generally negligible, so that these calculations are made as if the heat of combustion were not a function of temperature.

13.11 (a) $\Delta u = \Delta h - \Delta(Pv)$; For perfect gases $v = \dfrac{nRT}{P}$

$$\Delta u = \Delta h - \Delta\left[P\left(\frac{nRT}{P}\right)\right] = \Delta h - RT\Delta n$$

(b) For $\Delta n = 0, \quad \Delta u = \Delta h$

(c) For hydrocarbons $\Delta n = \left(x + \dfrac{y}{2}\right) - \left(1 + x + \dfrac{y}{4}\right) = \dfrac{y}{4} - 1$

For C_8H_{17}, $\quad \Delta n = \dfrac{17}{4} - 1 = 3.25 \dfrac{\text{mols}}{\text{mol fuel}}$

$$\Delta h - \Delta u = RT\Delta n = 1.987 \frac{\text{Btu}}{\text{lb mol}^\circ\text{R}} \cdot 528^\circ\text{R} \cdot 3.25 = 3410 \frac{\text{Btu}}{\text{lb mol}^\circ\text{R}} [@528^\circ\text{R}]$$

$$\frac{\Delta h - \Delta u}{\Delta u} = \frac{3410 \dfrac{\text{Btu}}{\text{lb mol}^\circ\text{R}}}{113 \dfrac{\text{lb}}{\text{lb mol}} \cdot 19{,}020 \dfrac{\text{Btu}}{\text{lb}}} = 0.0016$$

which shows that this difference is less than 0.1% for a typical fuel; it is commonly ignored.

13.12** (a) $T_2 = T_1 \left(\dfrac{V_1}{V_2}\right)^{\frac{R}{C_P - R}} = 4239^\circ\text{R} \cdot \left(\dfrac{1}{7}\right)^{\frac{1}{2.5}} = 1946^\circ\text{R} = 1486^\circ\text{F}$

(b) At idle (500 rpm) the time for heat transfer from the gases to the engine parts is four times as long as for typical roadway running (2000 rpm). Thus roughly four times as much heat is transferred from the gas. In addition, at idle the inlet pressure is much lower than during running, so that while the A/F is unchanged, the amount of fuel burned per combustion is less.

13.13 Non-Methane Organic Gases. Unfortunately the same materials are called HC (hydrocarbons) VOC (volatile organic compounds), NMHC (non-methane hydrocarbons), ROV (reactive organic vapors) and ROC (reactive organic compounds). There may be some other names I don't know about.

13.14 Here

$$K = \frac{[\mathrm{NO}]^2}{[\mathrm{N_2}]\cdot[\mathrm{O_2}]}; \ [\mathrm{O_2}] = \frac{[\mathrm{NO}]^2}{\mathrm{K}\cdot[\mathrm{N_2}]}$$

For 2560 K,

$$K = 21.9 \exp\left(\frac{-21{,}842}{2560}\right) = 4.315\,\mathrm{E}\text{-}3; \ (\text{See prob. } 12.8)$$

$$[\mathrm{O_2}] = \frac{[0.004]^2}{(4.315\,\mathrm{E}\text{-}3)\cdot(0.78)} = 0.00475$$

Repeating this calculation on a spreadsheet for the values listed one finds calculated oxygen concentrations between 0.5 and 0.4 %, average about 0.45%. The value is practically constant, and is plausible.

13.15 In that example the peak temperature was $2000 \text{ K} = 3140°\text{F}$. In this example the peak temperature $2355 \text{ K} = 3779°\text{F}$. At the higher temperature the reaction rate is high enough that the reverse reaction occurs, which is not the case in Figure 12.7. In Figure 13.3 the NO curve levels off at a crank angle of about 70° ATDC, which corresponds to a temperature of about 2100 K, very similar to the peak temperature in that example.

13.16 (a) $$\ln p\,(\text{psia}) = 11.724 - \left(\frac{5236.5}{560}\right); \quad p = 11.73\,\text{psia}$$

$$y_{\text{gasoline}} = \frac{10.73}{14.7} = 0.73; \ M_{\text{gasoline}} \approx 80\frac{\text{g}}{\text{mol}}(\text{assumed})$$

The assumed M is greater than in the examples in Chapter 10 because the percent vaporized is higher.

$$V_{\text{molar}} = 359.02\cdot\left(\frac{560}{492}\right) = 408.6\frac{\text{ft}^3}{\text{lb mol}} = 3056\frac{\text{gal}}{\text{lb mol}}$$

$$n_{\text{expelled}} = \frac{12 \text{ gal}}{3056\dfrac{\text{gal}}{\text{lb mol}}} = 0.00393 \text{ lb mol} = 1.78 \text{ mol}$$

$$\text{gasoline expelled} = 1.78 \text{ mol}\cdot 0.73\cdot 80\frac{\text{g}}{\text{mol}} = 104 \text{ g}$$

(b) The estimated capacity of the canister is 700 g·0.3 = 210 g. This is less than the value estimated above, suggesting that there is probably enough capacity. The current canister is designed for very low flow rates. It might need serious changes to accommodate the 10 gal/min of air expelled from the gasoline tank during fueling. The solution to the next problem suggests that even this need not be a serious problem.

13.17 (a) see pages 412 of *Fluid Mechanics for Chemical Engineers, 2e,* by Noel de Nevers, McGraw-Hill 1991. Assuming laminar flow, we can use the Kozeny-Carman equation. The particle diameter = 0.125 in = 3.17·10-3 m. If we take the mass of charcoal as the average of 700 and 800 = 750 g, then

$$\text{Bed volume} = \frac{750\text{ g}}{(30/62.4)\dfrac{\text{g}}{\text{L}}} = 1.56\text{ L} = \frac{\pi}{4}D^2L = 1.5\frac{\pi}{4}D^3$$

$$D = \left(\frac{4}{1.5\pi}\cdot V\right)^{1/3} = \left(\frac{4}{1.5\pi}\cdot 1560\text{ cm}^3\right)^{1/3} = 10.98\text{ cm}$$

$$L = 1.5D - 16.47\text{ cm}$$

Then

$$\Delta P = \frac{150\mu(1-\varepsilon)^2\Delta x}{D_p^2\varepsilon^3}\cdot V_s$$

$$= \frac{(150)\cdot\left(1.8\text{ E-5}\dfrac{\text{kg}}{\text{m}\cdot\text{s}}\right)\cdot(0.7)^2\cdot(0.1647\text{ m})}{(3.17\text{ E-3m})^2\cdot(0.3)^2}\cdot V_s = 562\frac{\text{Pa}}{\left(\dfrac{\text{m}}{\text{s}}\right)}\cdot V_s$$

(b) For 10 gal/min $V_s = \dfrac{\dot{Q}}{A} = \dfrac{10\dfrac{\text{gal}}{\text{min}}}{\dfrac{\pi}{4}(0.1098\text{ m})^2}\cdot\dfrac{0.003785\text{ m}^3}{\text{gal}}\cdot\dfrac{\text{min}}{60\text{ s}} = 0.066\dfrac{\text{m}}{\text{s}}$

so that $\Delta P = 560\dfrac{\text{Pa}}{\left(\dfrac{\text{m}}{\text{s}}\right)}\cdot 0.066\dfrac{\text{m}}{\text{s}} = 37\text{ Pa} = 0.148\text{ in H}_2\text{O}$

Which does not seem excessive.

13.18 (a) $\text{Area} = 1\text{ m}\cdot 0.02\text{ m}\cdot\pi = 0.00628\text{ m}^2$

$$\Delta\left(\begin{matrix}\text{HC}\\ \text{leakage}\end{matrix}\right) = (10-100)\frac{\text{g}}{\text{m}^2\text{day}}\cdot 0.00628\text{ m}^2 = -5.65\frac{\text{g}}{\text{day}}$$

(b) For 123 E6 cars and 365 days/yr

$$\dot{m} = 5.65\frac{\text{g}}{\text{day}}\cdot\frac{100}{90}\cdot 123\text{ E6}\cdot 365\cdot\frac{\text{tonne}}{1\text{ E6 g}} = 2.82\text{ E5}\frac{\text{tonne}}{\text{yr}}$$

$$= 2.82\text{ E5}\frac{\text{tonne}}{\text{yr}}\cdot\frac{1.1\text{ ton}}{\text{tonne}} = 3.1\text{ E5 }\frac{\text{ton}}{\text{yr}}$$

This is $\dfrac{3.1\text{ E5 } \dfrac{\text{ton}}{\text{yr}}}{19.2\text{ E6 } \dfrac{\text{ton}}{\text{yr}}} = 0.016 = 1.6\%$ of the total HC emissions for the US. When I first made up this problem I thought that using 365 days/yr was too high, since autos normally only run perhaps 2 hours per day. But the permeation continues all the time, if the hose is full of gasoline, which it normally is.

13.19 See the sketch at the right. The NO production requires that there be a high temperature and that there be enough oxygen. The dome is formed because one runs out of oxygen on the left of it and has too low a temperature for much NO formation at the right.

(I apologize for the mediocre sketch; in the next life I plan to become good at computer graphics!)

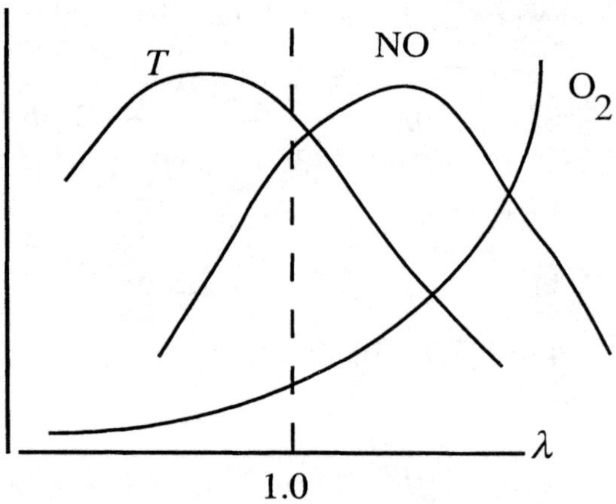

13.20 The combustion lasts long enough that the combustion gases do a substantial amount of work on the moving piston (and hence convert internal energy to work) before the combustion is complete. The peak temperature normally occurs shortly after TDC.

13.21 If one assumes sinusoidal motion of the piston then, maximum travel of piston = $2z$. At any instant the height of the piston above its lowest point = $z\ (1+\cos\theta)$

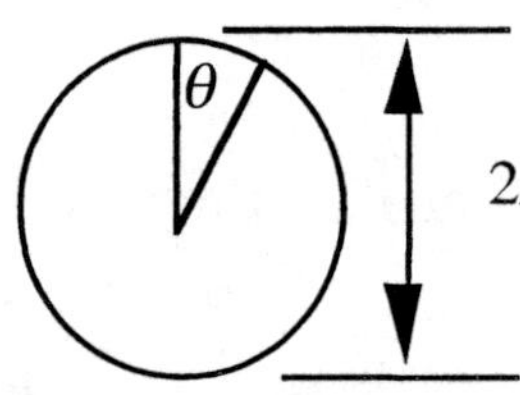

$$\frac{\text{Height}}{\text{Max height}} = \frac{z(1+\cos\theta)}{2z}$$

for $\theta = 0.15$ radians, this $= 0.983$

The actual piston path is more complex because while the crankshaft end of the connecting rod does undergo sinusoidal motion, the piston end does not. Taking this into account one finds

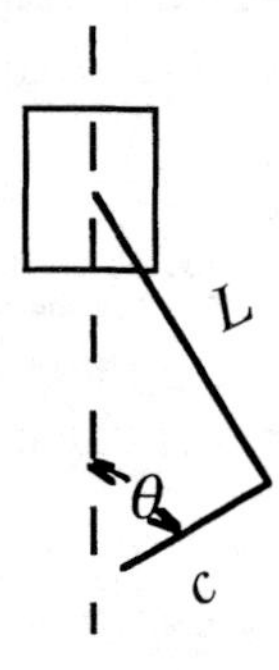

$$\frac{H}{H_{\max}} = 1 - \frac{1}{2}\left(1 + \frac{L}{c} - \cos\theta - \sqrt{\left(\frac{L}{c}\right)^2 - \sin^2\theta}\right)$$

where L is the length of the crankshaft and c is the radius of the crank. For $(L/c) = 2$ and 15° we have $\dfrac{H}{H_{\max}} = 0.9745$

13.22 (a) $$\text{Cost} = 1.5\text{ g} \cdot \frac{\$400}{\text{troy oz}} \cdot \frac{14.58\text{ troy oz}}{454\text{ g}} = \$19.30$$

I believe that there is a substantial industry recovering and reusing the precious metals from used automotive catalysts.

13.23 First estimate the maximum pressure in the cylinder

$$P_{max} = P_{atm} \cdot \frac{T}{T_o} \cdot \frac{V_o}{V} = 1\text{ atm}\left(\frac{4239°\text{R}}{528°\text{R}}\right)\frac{7}{1} = 56.2\text{ atm} = 827\text{ psia}$$

Then estimate the pressure of the exhaust

$$P_2 = P_1 \cdot \left(\frac{V_1}{V_2}\right)^k = 827\text{ psia}\left(\frac{1}{7}\right)^{1.4} = 53.4\text{ psia}$$

This is an overestimate for various reasons, but it shows that each opening of an exhaust valve is a "mini-explosion". The mufflers reduces that noise, at the expense of pressure drop and loss of fuel economy. The goal of muffler designers is to get maximum (noise reduction/loss of fuel economy), while keeping the size and cost of the muffler as low as possible (and providing a satisfactory service life).

13.24** From Figure 13.4, as we go from 15° after TDC to 75° after TDC, the temperature falls from about 2600 K to 2000 K, so that the rate of temperature decrease is about 10 K/ degree. One revolution = 360 degrees, and takes

$$\Delta t = \frac{\text{min}}{2000} \cdot \frac{60\text{ s}}{\text{min}} = 0.03\text{ s}$$

so that each degree corresponds to

$$\Delta t \text{ per degree} = \frac{0.03\text{ s}}{360} = 8.33\text{ E - 5}\frac{\text{s}}{\text{degree}}$$

and thus the rate of temperature decline is

$$\frac{\Delta T}{\Delta t} = \frac{-10\dfrac{\text{K}}{\text{degree}}}{8.33\text{ E - 5}\dfrac{\text{s}}{\text{degree}}} = -1.2\text{ E5}\frac{\text{K}}{\text{s}} = -2.16\text{ E5}\frac{°\text{F}}{\text{s}}$$

This is fast enough to "freeze" some chemical reactions, and not allow them to take up their equilibrium values at the lower temperatures reached.

Chapter 14

14.1 This is done on a spreadsheet. For $\lambda = 0.1$ microns, and 6000 K

$$E_{b\lambda}(\mathrm{T}) = \frac{374.15\ \mathrm{E}-18\ \mathrm{Wm}^2}{(1\,\mathrm{E}\text{-}7\mathrm{m})^5 \cdot \left(\exp\frac{14.388\ \mathrm{E}\text{-}3\ \mathrm{mK}}{(\mathrm{E}\text{-}7\mathrm{m})\cdot(6000\ \mathrm{K})} - 1\right)} = 1.441\ \mathrm{E}9\frac{\mathrm{W}}{\mathrm{m}^3} = 1.441\ \mathrm{E}3\frac{\mathrm{W}}{\mathrm{m}^2\mu}$$

By numerical integration, we have, from zero to 0.1 μ

$$\int_0^{0.1\mu} E_{\lambda b}d\lambda \cong E_{\lambda b}\cdot(0.1\mu - 0) = 1.441\,\mathrm{E}\text{-}2\frac{\mathrm{W}}{\mathrm{m}^2}$$

the sum of these numerical increments over the wavelength range from 0 to 10 μ is 6.71 E7 $\mathrm{W/m^2\mu}$. Dividing the above by this total value we find

$$\text{Normalized Fraction} = \frac{1.441\ \mathrm{E}2}{6.771\ \mathrm{E}7} = 2.12\ \mathrm{E}\text{-}6$$

which is the value plotted for 0.1 μ. The remaining values are found with a spreadsheet and plotted in that figure.

The same procedure is used for the earth, using $T = 288$ K.

14.2 Figure 14.5 shows (92+1) E9 tonne/yr going into the ocean and 90 E9 going back out. The rest is shown precipitated as carbonate on the ocean floor, and is not a cause for any concern. Most proposals for reducing the CO_2 content of the atmosphere suggest that we should reduce emissions. Some suggest that we should speed the rate at which the ocean removes CO_2 and deposits it as limestone.

The principal purpose of this problem and the next is to make the students study that figure.

14.3 The figure shows the inventory of carbon dioxide in the atmosphere as 750, and the flows in and out as about 150 E9 tonne/yr. Thus the average time a molecule spends must be (750/150) = about 5 years. This is cited by many who feel that our concern with the greenhouse effect is unwarranted, to show that if we find that such warming is occurring we can deal with it quickly. Those who believe that the greenhouse effect is a major problem reply that although the residence time of a CO_2 molecule in the atmosphere is short, that in the oceans is 40,000/150 267 years, and that the ocean holds (40,000/750) 53 times as much as the atmosphere. Thus if we raise the CO_2 content of the atmosphere we will also raise that of the ocean, and that while the ocean is slow to have its CO_2 concentration increase, it will also be slow to have it decrease if we decide we must do that. This question is one of current scientific debate (see Ref 26).

14.4 (a) The only alkali available on that scale is lime or limestone. However to make it alkaline CO_2 must be driven off. If we wanted to do that, we would want to produce the lime in plants where the CO_2 was produced in concentrated form, and then disposed of some way (old natural gas fields? dissolved in the deep ocean?)

(b) That is an interesting short term solution. However once the new vegetative cover is established, somewhere where there is none, thus removing a suitable amount of CO_2 from the atmosphere, the vegetation at steady state will return as much CO_2 to the atmosphere as it removes. If the vegetation is harvested and burned for fuel, thereby replacing an equivalent mass of fossil fuels, then this would have positive effects on the global CO_2 balance.

14.5 Whoever designed the eye knew the wavelength range where the most light was. One can assume that this was simple natural selection, or make some other assumption involving divine intervention. Some insects can see ultraviolet light. As far as we know they developed their vision system independent of ours. As far as we know, many mammals are color blind. Humans are a very visual species, compared, for example to dogs who do a lot more by smell. That may explain why we developed color vision, which many animals did not.

14.6** $\Delta x = x \cdot coef \cdot \Delta T = (1000\text{m}) \cdot \left(\frac{0.00012}{°\text{C}}\right) \cdot (1°\text{C}) = 12\,\text{cm} = 4.7\,\text{inches}$

This doesn't seem like much, but it makes a difference in low lying areas, like Holland, the coral islands of the Pacific or Bangladesh.

14.7 (a) $EF = \dfrac{5.2\text{ E6 tons}}{120{,}000\text{ acres} \cdot 40\text{ years}} = 1.08\dfrac{\text{ton CO}_2}{\text{acre} \cdot \text{year}}$

(b) On a carbon basis this is $EF = 1.08\dfrac{\text{ton CO}_2}{\text{acre} \cdot \text{year}} \cdot \dfrac{12\text{ ton C}}{44\text{ ton CO}_2} = 0.295\dfrac{\text{ton C}}{\text{acre} \cdot \text{yr}}$ wood is 50 wt % carbon, so that

$$\text{Wood production} = 0.295\frac{\text{ton C}}{\text{acre} \cdot \text{yr}} \cdot \frac{1}{0.5} = 0.59\frac{\text{ton wood}}{\text{acre} \cdot \text{yr}} = 1181\frac{\text{lb wood}}{\text{acre} \cdot \text{yr}}$$

(c) $1181\dfrac{\text{lb wood}}{\text{acre} \cdot \text{yr}} \cdot \dfrac{\text{acre}}{400\text{ trees}} = 2.95\dfrac{\text{lb}}{\text{tree} \cdot \text{yr}}$ which does not seem like a high estimate.

(d)

$$\begin{pmatrix}\text{annual}\\ \text{electricty}\\ \text{production}\end{pmatrix} = 840\text{ MW} \cdot 0.7 \cdot 365 \cdot 24\frac{\text{hr}}{\text{yr}} = 5.15\text{ E6}\frac{\text{MWhr}}{\text{yr}}$$

$$\begin{pmatrix}\text{annual}\\ \text{coal use}\end{pmatrix} = 5.15\text{ E6}\frac{\text{MWhr}}{\text{yr}} \cdot \frac{1}{0.35} \cdot \frac{\text{Btu}}{2.93\text{ E-4 kWh}} \cdot \frac{\text{lb coal}}{13{,}600\text{ Btu}} = 3.68\text{ E9}\frac{\text{lb}}{\text{yr}}$$

$$\begin{pmatrix}\text{annual}\\ \text{CO}_2\text{ production}\end{pmatrix} = 3.68\text{ E9}\frac{\text{lb coal}}{\text{yr}} \cdot 0.758\frac{\text{lb C}}{\text{lb coal}} \cdot \frac{44\text{ lb CO}_2}{12\text{ lb C}} \cdot \frac{\text{ton}}{2000\text{ lb}}$$

$$= 5.12\text{ E6}\frac{\text{ton}}{\text{yr}}$$

and for 40 years 205 million tons of CO_2 or about 39 times the amount to be sequestered by the forest preserve. This is not to suggest that the forest preserve is a bad idea, but only to show the magnitudes of the CO_2 flows involved.

14.8 See Example 14.2

$$T^4 = \frac{(288.15\text{ K})^4}{0.9};\ T = \frac{(288.15\text{ K})}{0.9^{1/4}} = 295.94\text{ K}$$

$$\Delta T = 295.84 - 288.15 = 7.69\text{ K}$$

No one has suggested that we could make this large an increase in the absorptivity of the atmosphere for outgoing terrestrial radiation in any reasonable time period, so this scary number is merely an illustration of what would happen if we did. This is apparently what happened on Venus, where the atmosphere has enough IR absorbers that its temperature is much, much higher than it would be if it had the same atmosphere we do.

14.9 (a) This topic is discussed in Section 19 of *Marks' Standard Handbook for Mechanical Engineers* Eighth Edition, McGraw-Hill, 1978

(b) Current efforts are directed at HCFCs, which contain at least one hydrogen, and thus can be attacked by OH radical in the atmosphere and at fluorocarbons which contain no chlorine and are apparently not damaging to the ozone layer. Some mixtures of such compounds have properties similar to the Freons. None is apparently quite as good, but they are close.

The refrigerant suppliers (e.g. DuPont) are developing and supplying mixtures of these materials (plus hydrocarbons like propane) whose mixture properties are very close to those of the common chlorofluorocarbon refrigerants, so that they can be used satisfactorily in existing equipment which was designed to use chlorofluorocarbons. Du Pont sells these under the trade name "SUVA". If you call them at 1-800-235-SUVA they will send you a huge package of technical data on them.

14.10** (a) $@\ \text{pH} = 3.8,\ [\text{H}^+] = 10^{-3.8} = 1.58\text{ E-}4\frac{\text{mol}}{\text{L}}$

$$@\ \text{pH} = 2.8,\ [\text{H}^+] = 1.58\text{ E-}3\ \frac{\text{mol}}{\text{L}}$$

$$\Delta[H^+] = 1.58\text{ E-}3 - 1.58\text{E-}4 = 1.42\text{E-}3\frac{\text{mol}}{\text{L}}$$

$$\text{For H}_2\text{SO}_4\text{, 1 mol of H}^+ = \frac{98}{2} = 49\text{ g,}$$

$$\text{So we need } 49\frac{\text{g}}{\text{mol H}^+}\cdot 1.42\text{ E-}3\frac{\text{mol H}^+}{\text{L}} = 0.0696\frac{\text{g}}{\text{L}}$$

$$\rho_{\text{H}_2\text{SO}_4} = 1.38\frac{\text{g}}{\text{cc}};\ v = \frac{0.0696\ \frac{\text{g}}{\text{L}}}{1.83\frac{\text{g}}{\text{cc}}} = 0.038\frac{\text{cc}}{\text{L}}$$

This is slightly less than one drop from a standard laboratory buret per liter of wine

(b) At equilibrium there will be zero H_2SO_4 and we will have the ratio

$$\frac{[SO_4^{=}]}{[HSO_3^{-}]} = \frac{K_2}{[H^+]} = \frac{0.012}{1.58\text{ E - 3}} = 7.595$$

The final hydrogen ion concentration is equal to the initial concentration plus the amount contributed by the sulfuric acid, or

$$[H^+] = 1.58\text{ E - 3} = 1.58\text{E - 4} + 2[SO_4^{=}] + [HSO_4^{-}] = 1.58\text{ E - 4} + [SO_4^{=}] \cdot \left\{1 + \frac{1}{7.595}\right\}$$

the concentration of SO_4^{-2} is

$$[SO_4^{=}] = \frac{1.58\text{ E - 3} - 1.58\text{ E - 3}}{2.1317} = 6.6707\text{ E - 4}\frac{\text{mol}}{\text{L}}$$

and that of HSO_4^- is

$$[HSO_4^{-}] = \frac{6.6707\text{ E - 4}}{7.595} = 8.78\text{ E - 5}\frac{\text{mol}}{\text{L}}$$

so that $[SO_4^{=}] + [HSO_4^{-}] = 6.6707\text{ E} - 4 + 8.78\text{E} - 5 = 7.55\text{ E} - 4\frac{\text{mol}}{\text{L}}$

$$7.55\text{ E - 4}\frac{\text{mol}}{\text{L}} \cdot \frac{98\text{g}}{\text{mol}} = 0.074\frac{\text{g}}{\text{L}}$$

This is (0.074/0.0696) = 1.06 times the amount required in part (a), because only 7/8 of the HSO_3^- contributes an H^+.

This problem exposes the students to acidity calculations, and shows that the amounts of acid needed to make significant changes in pH are quite small.

While it is always a pleasure to make fun of wine snobs, and particularly of French wine snobs, this is an example of a law which has no scientific basis, and whose violation harms no one, except those who make a living by wine snobbery. In the USA sulfuric acid is on the GRAS (Generally Recognized As Safe) list of chemicals which may be added to foodstuffs without testing.

14.11 For a "perfect acid" a 1 N solution would have pH = 0. All real acids are not perfect in the sense that they are not completely ionized so that a 1 N solution does not produce 1 N of H^+ ions. If the reported value for HCl is correct, then one may show that its ionization constant is about 8.6. If one uses the methods in problem 14.10 (b) to compute the pH of 1 N H_2SO_4 one finds pH = 0.29.

14.12** Total water $= 1 \text{ m} \cdot 10 \text{ km}^2 \cdot 1\text{ E6} \dfrac{\text{m}^2}{\text{km}^2} = 1\text{ E7m}^3 = 1\text{ E10 L}$

$$\Delta[\text{H}^+] = 10^{-4..5} - 10^{-5.6} = 31.6\text{ E-6} - 2.51\text{ E-6}\frac{\text{mol}}{\text{L}} = 29.1\text{ E-6}\frac{\text{mol}}{\text{L}}$$

The lime must neutralize $1\text{ E10 L} \cdot 2.91\text{ E-5}\dfrac{\text{mol}}{\text{L}} = 2.91\text{ E5 mol}$ of H^+. $CaCO_3$ neutralizes 2 mols of H^+ per mol, so we need $\dfrac{100\text{ g}}{2\text{ mol}} \cdot 2.91\text{ E5 mol} = 1.46\text{ E7g} = 14.6$ tonne or about one truckload per year. This is widely practiced in Sweden. The lime is finely ground and slurried with water so that it can be well dispersed in the lake.

When I gave this as a homework problem most of the students looked at Equation 14.11 and concluded that one mole of limestone would only remove one mol of H^+. That is not correct. There is the additional reaction

$$HCO_3^- + H^+ \Leftrightarrow H_2CO_3 \Leftrightarrow H_2O + CO_2$$

that removes another mole of H^+ per mole of limestone, releasing CO_2 gas to the atmosphere.

The equilibrium in this reaction , at room temperature, is given by

$$\frac{[\text{H}_2\text{CO}_3]}{[\text{H}^+]\cdot[\text{HCO}_3^-]} = \frac{1}{K_{20°\text{C}}} = \frac{1}{4.3\text{ E-7 mol/L}}$$

The concentration of H_2CO_3 in water in equilibrium with the atmosphere (at 20°C) is about 1.08·E-5 mol/L. For this example we can compute that if this reaction did not occur, the concentration of H_2CO_3 would be

$$[\text{H}_2\text{CO}_3] = \frac{[\text{H}^+]\cdot[\text{HCO}_3^-]}{4.3\text{ E-7}} = 1.69\text{ E-4}\frac{\text{mol}}{\text{L}}$$

which is about 15 times the equilibrium value. Thus this reaction does occur, and we remove 2 moles of H^+ per mole of Ca, as the above solution shows.

14.13

$$\text{If } [\text{H}^+] = 8.25\text{ E-5}\frac{\text{mol}}{\text{L}}, \quad \text{pH} = -\log_{10}(8.25\text{ E-5}) = 4.083$$

vs 4.071 in the example. This is a consequence of the logarithmic definition of pH.

Chapter 15

15.1 (a) One has to go back to the original to see all the manipulations, but this is a material balance, in which the first term represents the increase with time in CO in the system, the second is the rate of production of CO in the system the third is the rate of removal of CO by being breathed out and the fourth is the amount of addition of CO by being breathed in.

The authors use a mass transfer equation with D_L as (mass transfer coefficient·area) to eliminate the unknown concentration of CO in the lungs.

(b) Making the A, B and P_L substitutions and dividing by V_b

$$\frac{d[\text{COHb}]}{dt} = \frac{\dot{V}_{\text{CO}}}{V_b} - \frac{A}{BV_b}[\text{COHb}] + \frac{P_{\text{I}_{\text{CO}}}}{V_b B}$$

$$\int \frac{d[\text{COHb}]}{[\text{COHb}] - \frac{B}{A}\dot{V}_{\text{CO}} - \frac{P_{\text{IC}_{\text{CO}}}}{A}} = \int \frac{-A}{V_b B} dt;$$

$$\ln\left\{[\text{COHb}] - \frac{B}{A}\dot{V}_{\text{CO}} - \frac{P_{\text{I}_{\text{CO}}}}{A}\right\}\Bigg|_0^t = \frac{-At}{V_b B}\Bigg|_0^t$$

$$\frac{A[\text{COHb}]_t - B\dot{V}_{\text{CO}} - P_{\text{ICO}}}{A[\text{COHb}]_0 - B\dot{V}_{\text{CO}} - P_{\text{ICO}}} = \exp\frac{-At}{V_b B} 1.46\text{ E7g} = 14.6 \text{ tonne}$$

15.2** (a) When $t = \infty$, exp (- etc.·t) = 0 so that $[\text{COHb}]_\infty = \frac{B}{A}\dot{V}_{\text{CO}} + \frac{P_{\text{I}_{\text{CO}}}}{A}$

(b) For 100 ppm , (using the values on Figure 15.1) we have

$$A = \frac{\overline{P}_{CO_2}}{M[\text{O}_2\text{Hb}]} = \frac{100 \text{ mm Hg}}{218 \cdot \left[0.20 \frac{\text{ml(stp)}}{\text{ml}}\right]} = 2.29 \frac{\text{mm Hg}}{\frac{\text{ml(stp)}}{\text{ml}}}$$

$$B = \frac{1}{D_L} + \frac{P_L}{\dot{V}_A} = \frac{1}{30\frac{\text{ml}}{\text{mm Hg}}} + \frac{707 \text{ mm Hg}}{6000\frac{\text{ml}}{\text{min}}} = 0.1512 \frac{\text{min} \cdot \text{mm Hg}}{\text{ml}}$$

$$\frac{B}{A}\dot{V}_{\text{CO}} = \frac{0.1512\frac{\text{min} \cdot \text{mm Hg}}{\text{ml}}}{2.29\frac{\text{mm Hg} \cdot \text{ml}}{\text{ml(stp)}}} \cdot 0.007 \frac{\text{ml}}{\text{min}} = 4.62\text{ E - 4} \frac{\text{ml(stp)}}{\text{mm}}$$

$$\frac{P_{\text{I}_{\text{CO}}}}{A} = \frac{(750 \text{ mm Hg} \cdot 1\text{ E - -4})}{2.29\frac{\text{mm Hg} \cdot \text{ml}}{\text{ml(stp)}}} = 0.0327 \frac{\text{ml(stp)}}{\text{ml}}$$

$$[\text{COHb}] = 0.00046 + 0.0327 = 0.0332\frac{\text{ml(stp)}}{\text{ml}};\quad \%\text{COHb} = \frac{0.0332}{0.200}\cdot 100 = 16.6\%$$

The value from the plot, as best I can read it, is 15%. For 25 and 50 ppm the results of the above calculation (done on a spreadsheet) agree with the plot within chart reading accuracy. For 1000 ppm the above calculation leads to 164%, which is clearly wrong, as discussed in the problem statement.

(c) The first term in the above sum is the contribution of internally-generated CO, leading to

$$\%\text{CO Hb} = \frac{0.00046}{0.20}\cdot 100 = 0.23\%$$

15.3 (a) $$[\text{CO Hb}]_t = \frac{B}{A}\dot{V}_{\text{CO}} + \frac{P_{I\text{CO}}}{A} + \left[[\text{CO Hb}]_0 - \frac{B}{A}\dot{V}_{\text{CO}} - \frac{P_{I\text{CO}}}{A}\right]\cdot \exp\frac{-At}{V_b B}$$

From Prob. 15.2; $\left(\frac{B}{A}\dot{V}_{\text{CO}} + \frac{P_{I\text{CO}}}{A}\right) = 0.0332\frac{\text{ml (stp)}}{\text{ml}}$

$$[\text{CO Hb}]_0 = 0.008\cdot 0.20 = 0.0016\frac{\text{ml (stp)}}{\text{ml}}$$

$$\frac{At}{V_b B} = \frac{2.29\dfrac{\text{mm Hg}}{\dfrac{\text{ml(stp)}}{\text{ml}}}\cdot 100\,\text{min}}{5500\,\text{ml}\cdot 0.1512\dfrac{\text{min mm Hg}}{\text{ml}}} = 0.275$$

$$[\text{CO Hb}]_{100\,\text{min}} = 0.0332 + [0.0016 - 0.0332]\cdot \exp(-0.275) = 0.0092\frac{\text{ml(stp)}}{\text{ml}}$$

$\frac{0.0092}{0.20} = 0.046 \approx 5\%$ The plot shows 5%

(b) For 70% after five minutes

$$\frac{At}{V_b B} = 0.275\cdot\frac{5}{100} = 0.01375;\quad \exp(-0.01375) = 0.9863$$

The only terms that play a role are

$$[\text{CO Hb}] - \frac{P_{I\text{CO}}}{A} = -\left(\frac{P_{I\text{CO}}}{A}\right)\cdot(0.9863);\quad [\text{CO Hb}] = \frac{P_{I\text{CO}}}{A}\cdot(1 - 0.9863)$$

$$P_{I\text{CO}} = \frac{[\text{CO Hb}]\cdot A}{0.0137} = \frac{\left(0.7\cdot 0.2\dfrac{\text{ml(stp)}}{\text{ml}}\right)\left(2.29\dfrac{\text{ml Hg}}{\text{ml(stp)/ ml}}\right)}{0.0137} = 23.4\ \text{mm Hg}$$

$$y_{\text{CO}} = \frac{23.4\ \text{mm Hg}}{760\ \text{mm Hg}} = 0.0301 \approx 3\% = 30{,}000\ \text{ppm}$$

Which is 20 times the IDLH value. OSHA defines ".. (IDLH) represents the maximum concentration from which, in the event of respirator failure, one could escape within 30 minutes without a respirator and without experiencing any escape-impairing effects (e.g. severe eye irritation) or irreversible health effects". Thus IDLH is not expected to be as severe as "kills in five minutes". The OSHA TWA for CO is 35 ppm.

15.4 (a) Airflow = $Q_{air} = 1.8\,\mathrm{L}\cdot\frac{715}{\mathrm{min}}\cdot\frac{1}{2} = 643\frac{\mathrm{L}}{\mathrm{min}} = 38{,}500\frac{\mathrm{L}}{\mathrm{hr}}$

$$\dot{n}_{air} = 643\frac{\mathrm{L}}{\mathrm{min}}\cdot\frac{\mathrm{mol}}{22.4\,\mathrm{L}}\cdot\frac{273\,\mathrm{K}}{293\,\mathrm{K}} = 26.7\frac{\mathrm{mol}}{\mathrm{min}} = 1602\frac{\mathrm{mol}}{\mathrm{hr}}$$

$$\dot{n}_{\mathrm{CO}} = 26.7\frac{\mathrm{mol}}{\mathrm{min}}\cdot 300\,\mathrm{E}\text{-}6 = 0.0080\frac{\mathrm{mol}}{\mathrm{min}}$$

$$\dot{m}_{\mathrm{CO}} = 0.0080\frac{\mathrm{mol}}{\mathrm{min}}\cdot 28\frac{\mathrm{g}}{\mathrm{mol}} = 0.22\frac{\mathrm{g}}{\mathrm{min}} = 13.2\frac{\mathrm{g}}{\mathrm{hr}}$$

(b) Oxygen consumption $= 26.7\frac{\mathrm{mol}}{\mathrm{min}}\cdot 0.21 = 5.6\frac{\mathrm{mol}}{\mathrm{hr}} = 336\frac{\mathrm{mol}}{\mathrm{hr}} = 9744\frac{\mathrm{g}}{\mathrm{hr}}$

(c) Fresh air flow $= 20\cdot 20\cdot 8\frac{\mathrm{ft}^3}{\mathrm{hr}}\cdot\frac{\mathrm{lbmol}}{385\,\mathrm{ft}^3} = 8.31\frac{\mathrm{lbmol}}{\mathrm{hr}} = 3770\frac{\mathrm{mol}}{\mathrm{hr}} = 63\frac{\mathrm{mol}}{\mathrm{min}}$

$$c_{\mathrm{CO}} = \frac{\dot{n}_{\mathrm{CO}}}{\dot{n}_{air}} = \frac{0.0080\frac{\mathrm{mol}}{\mathrm{min}}}{63\frac{\mathrm{mol}}{\mathrm{min}}} = 1.27\mathrm{E}-4 = 127\mathrm{ppm};\quad c_{\mathrm{O}_2} = 0.21 - \frac{5.6}{63} = 0.12 = 12\%$$

(d) The CO concentration is most likely not fatal. The low O_2 concentration may be more dangerous, but it also may not be fatal. If you take off or disable your catalytic converter, then the CO concentration in the garage will be roughly 30 times as large, and almost certainly fatal. I repeat, don't try this at home to see if I am right. I won't.

15.5 Solving Equation 5 for $P_{\mathrm{I_{CO}}}$ and then observing that the internal generation term is negligible, we find

$$P_{\mathrm{I_{CO}}} = B\dot{V}_{CO} - A[\mathrm{COHb}]_0\cdot\left\{\frac{\exp\left(\frac{-tA}{V_bB}\right)}{1-\exp\left(\frac{-tA}{V_bB}\right)}\right\} + \frac{A[\mathrm{COHb}]_t - B\dot{V}_{CO}}{1-\exp\left(\frac{-tA}{V_bB}\right)}$$

Then using the values of A and B from the previous few problems we construct the curve for [COHb] = 10 % as follows

$$B\dot{V}_{CO}\approx 0;\ [\mathrm{COHb}]_0\approx 0;\ P_{\mathrm{I_{CO}}}\approx\frac{A[\mathrm{COHb}]}{\left(1-\exp\left(\frac{-tA}{V_bB}\right)\right)}$$

$$P_{I_{CO}} = 2.29\frac{\text{mm Hg}}{\frac{\text{ml(stp)}}{\text{ml}}} \cdot \frac{\left[0.1 \cdot 0.20\frac{\text{ml(stp)}}{\text{ml}}\right]}{\left(1-\exp\left(\frac{-0.00275}{\text{mm}} \cdot t\right)\right)}$$

For 100 minutes this becomes

$$P_{I_{CO}} = \frac{(2.29)\cdot(0.02)}{1-\exp(-0.00275\cdot 100)} = 0.190\,\text{mm Hg}$$

$$y_{CO} = \frac{0.190\,\text{mm Hg}}{760\,\text{mm Hg}} = 2.5\,\text{E-4} = 250\,\text{ppm}$$

which is equal to the value on Figure 15.1 (within chart reading accuracy). Then, using a spreadsheet one computes the values for various times, and plots them on a log-log scale, finding a curve (for 10% COHb) of the same shape as those in Figure 2.5.

15.6** $\log_{10} p = A - \frac{B}{(T+C)}; \quad T + C = \frac{B}{A - \log_{10} p}$

For 10 mm Hg

$$T + C = \frac{9.845.4}{(7.829 - \log_{10} 10)} = 1442.13\,\text{K} = 2597^{\circ}\text{R} = 2136^{\circ}\text{F}$$

and for 100 mm Hg, $T = 2581°F$. These temperatures are regularly observed in incinerators and coal combustion, so that elemental lead would be vaporized to a significant extent. In the cooler parts of the furnace it would be expected to condense into fine particles which would escape if the particle collection device were not of high quality. Some lead compounds have boiling points much lower than that of metallic lead, e.g. lead chloride boils at 1749°F.

15.7 (a) $EF = 0.065 \text{ to } 0.09\frac{\text{kg}}{\text{tonne}} = (0.065 \text{ to } 0.09)\cdot\frac{\text{tonne}}{1000\text{ kg}}$

$= 0.0065 \text{ to } 0.009\,\text{wt\%} = 65 \text{ to } 90 \text{ ppm}$

(b) Most solders have lead. That includes the solders in electronic equipment, some tin cans, old air conditioners and refrigerators. Lead is used for many seals, e.g. the caps on wine bottles. (This is foolish snobbery. There are other kinds of seals which are just as effective and cheaper and not a problem in incinerators, but a high priced wine must have the traditional lead seal, etc.)

If the lead forms an oxide in the incinerator, then it is likely to become a solid particle before reaching the particle collection device. If it forms a chloride, which has a much lower melting (or boiling) point temperature, then it is likely to pass through the particle collector in the gas phase. There is ample chlorine in PVC plastic in most municipal waste.

15.8 For Ex 15.1 $\quad Q_{\text{fan}} = 200\text{ cfm} \cdot \dfrac{60\text{ min}}{\text{hr}} = 12{,}000\dfrac{\text{ft}^3}{\text{hr}}$

$$\text{For 1 hr,}\quad c_1 = 0.20\frac{\mu\text{g}}{\text{m}^3} + \frac{0.183\frac{\text{g}}{\text{hr}}}{12{,}000\frac{\text{ft}^3}{\text{hr}}} \cdot 35.29\frac{\text{ft}^3}{\text{m}^3} = 20 + 538 = 558\frac{\mu\text{g}}{\text{m}^2}$$

$$\text{For 24 hr,}\quad c_{24} = 0.20 + \frac{538}{24} = 20 + 22.4 = 44.4\frac{\mu\text{g}}{\text{m}^3}$$

and for Ex. 15.2 $\quad (c_o - c_i)_{\text{initial}} = 558\dfrac{\mu\text{g}}{\text{m}^3};\ \dfrac{Q}{V} = \dfrac{12{,}000}{1{,}000} = \dfrac{12}{\text{hr}}$

$$c = 20\frac{\mu\text{g}}{\text{m}^3} + 558\frac{\mu\text{g}}{\text{m}^3}\cdot\left(\exp\frac{-12}{\text{hr}}\cdot 1\,\text{hr}\right) = 20 + 0.003 = 20.003\frac{\mu\text{g}}{\text{m}^3}$$

15.9 Starting with Ex. 15.3 $\quad V\dfrac{dc_i}{dt} = Q_1 c_o - Q_2 c_i + S;\quad \dfrac{dc_i}{c_o - c_i + \dfrac{S}{Q}} = -\dfrac{Q}{V}dt$

$$\ln\left(c_i - c_o - \frac{S}{Q}\right)\Bigg|_s^f = \frac{-Q}{V}t\Bigg|_s^f;\quad \ln\left\{\frac{c_i - c_o - \dfrac{S}{Q}}{-\dfrac{S}{Q}}\right\} = \frac{-Q}{V}\Delta t$$

$$c_i = c_o + \frac{S}{Q}\cdot\left(1 - \exp-\frac{Q\Delta t}{V}\right) = 20\frac{\mu\text{g}}{\text{m}^3} + 2153\frac{\mu\text{g}}{\text{m}^3}\cdot\left(1 - \exp-\frac{3000\frac{\text{ft}^3}{\text{hr}}}{1000\,\text{ft}^3}\cdot 0.25\,\text{hr}\right)$$

$$= 20 + 2153\cdot(0.528) = 1156\frac{\mu\text{g}}{\text{m}^3}$$

15.10 Step 1, see Prob. 15.8

$$c_i = c_o + \frac{S}{Q}\cdot\left(1 - \exp-\frac{Q\Delta t}{V}\right) = 20\frac{\mu\text{g}}{\text{m}^3} + 2153\frac{\mu\text{g}}{\text{m}^3}\cdot\left(1 - \exp-\frac{3000\frac{\text{ft}^3}{\text{hr}}}{1000\,\text{ft}^3}\cdot 0.5\,\text{hr}\right) = 1693\frac{\mu\text{g}}{\text{m}^3}$$

Step 2, see Ex. 15.3

$$c_i = c_o - (c_o - c_i)\cdot\exp-\frac{Q\Delta t}{V} = 20\frac{\mu\text{g}}{\text{m}^3} + (1693 - 20)\frac{\mu\text{g}}{\text{m}^3}\cdot\exp-\frac{3000\frac{\text{ft}^3}{\text{hr}}}{1000\,\text{ft}^3}\cdot 0.25\,\text{hr} = 810\frac{\mu\text{g}}{\text{m}^3}$$

Step 3, start with Eq. 15.3

$$V\frac{dc_i}{dt} = Q_1 c_o - Q_5 c_i + S;\quad \frac{dc_i}{\frac{1}{V}\cdot(Q_1 c_o - Q_5 c_i + S)} = dt;\qquad \ln\frac{c_{i\,final} - c_o - \dfrac{S}{Q}}{c_{i\,initial} - c_o - \dfrac{S}{Q}} = -\frac{Q}{V}\Delta t$$

$$c_{i\ final} = c_o + \frac{S}{Q} + \left(c_{i\ initial} - c_o - \frac{S}{Q} \right) \cdot \exp\left(-\frac{Q}{V} \Delta t \right)$$

$$c_{i\ final} = 20 \frac{\mu g}{m^3} + 2154 \frac{\mu g}{m^3} + (810 - 20 - 2154) \frac{\mu g}{m^3} \cdot \exp\left(- \frac{3000 \frac{ft^3}{hr}}{1000 \frac{ft^3}{hr}} \cdot 0.25\ hr \right) = 1530 \frac{\mu g}{m^3}$$

We conclude that for this somewhat artificial problem, the peak concentration is at the end of the first cooking period.

15.11 (a) For steady state and recirculation only

$$0 = Q_4 c_i (1 - F_0) - Q_4 c_i + S - R; \quad c_i = \frac{S - R}{Q_4 F_0}$$

(b) $$S = 100 \cdot 0.63 \frac{scf}{hr} \cdot \frac{44\ lb}{386\ scf} = 7.2 \frac{lb}{hr}$$

The air in submarines is circulated over some type of absorbents, presumably alkaline. In the movie "Apollo 13" the CO_2 buildup in the cabin was one of the dramatic plot turns; there the CO_2 was removed by solid adsorbents.

(c) In houses we count on open windows and/or the unintended air infiltration from outside to dilute and remove the odorous compounds, or if that is deemed inadequate, we use an "air freshener" which emits a stronger odor to cover over the unpleasant smells. In commercial buildings we use fresh makeup air to dilute and remove the odorous compounds. In submarines we adsorb the odorous compounds on activated charcoal beds, through which the submarine's air is circulated.

15.12 (a) $$S = \frac{300\ kcal}{hr} \cdot \frac{645\ \mu g}{kcal} = 1.93 \frac{g}{hr}$$

$$c_o = 1\ ppm = (1\,E\text{-}6) \cdot \left(\frac{28}{29} \right) \cdot \left(1.20 \frac{kg}{m^3} \right) \cdot \left(\frac{1000 g}{kg} \right) = 1158 \frac{\mu g}{m^3}$$

$$c_i = 1158 \frac{\mu g}{m^3} + \frac{1.93 \frac{g}{hr}}{3000 \frac{ft^3}{hr}} \cdot \frac{35.29\ ft^3}{m^3} = 1158 + 22{,}703 = 23{,}861 \frac{\mu g}{m^3}$$

$$c_{avg} = 1158 + \frac{22{,}703}{24} = 2104 \frac{\mu g}{m^3}$$

(b) The peak value is 22/40 = 58% of the one hour CO standard. For NO_2 the peak value is 110/100 = 110 % of the **annual** average NO_2 standard.

15.13 By reading values from the figure and taking the ratio (on a spreadsheet) one finds that the kitchen value is on the average (2.24 · the bedroom value), with a standard deviation of 0.32. This suggests the kind of model sketched in Figure 15.5.

15.14 (a) For the kitchen $(Q_1 + Q_5)c_k = Q_1 c_o + Q_5 c_r + S$

For the rest of the house $(Q_1 + Q_5)c_r = Q_1 c_o + Q_5 c_k$

Eliminating c_r between these two equations and simplifying we find

$$c_k = \frac{c_o \cdot \left(Q_1 + \dfrac{Q_3 Q_5}{Q_4 + Q_6}\right) + S}{(Q_1 + Q_4) - \dfrac{Q_5^{\ 2}}{Q_4 + Q_5}}$$

Next we convert the emission rate and outdoor concentration to ft^3, and the emission rate to μg, $c_o = 20\dfrac{\mu g}{m^3} \cdot \dfrac{m^3}{35.32\ ft^3} = 0.566\dfrac{\mu g}{ft^3}$; $S = 0.183\dfrac{g}{hr} = 183{,}000\dfrac{\mu g}{hr}$

$$c_k = \frac{0.566\dfrac{\mu g}{ft^3} \cdot \left(1000 + \dfrac{10{,}000 \cdot 10{,}000}{10{,}000 + 10{,}000}\right)\dfrac{ft^3}{hr} + 183{,}000\dfrac{\mu g}{hr}}{\left(1000 + 10{,}000 - \dfrac{10{,}000 \cdot 10{,}000}{10{,}000 + 10{,}000}\right)\dfrac{ft^3}{hr}} = 31.07\frac{\mu g}{ft^3} = 1097\frac{\mu g}{m^3}$$

Then we solve for c_r

$$c_r = \frac{Q_3 c_o + Q_4 c_k}{Q_6 + Q_4} = 0.5c_o + 0.5c_k = 10 + 1097 = 558.6\frac{\mu g}{m^3}$$

(b) The rigorous solution requires the simultaneous solution of the following two ODEs:

$$V_k \frac{dc_k}{dt} = Q_1 c_o + Q_5 c_r - (Q_2 + Q_5)c_r + S$$

$$V_r \frac{dc_r}{dt} = Q_3 c_o + Q_6 c_k - Q_5 c_r$$

which can certainly be solved numerically, for each of the time periods. One may make a fair estimate by noting that only the average value is called for, and that the stove only runs 2/24 of the time, so that average emission rate is 1/12 of the values shown above, leading to 109.8 and 64.9 μg/m³.

See also the solution to the next problem.

15.15 For the kitchen $V_k \dfrac{dc_k}{dt} = Q_1 c_o + Q_3 c_r - (Q_1 + Q_3)c_k + S$

For the rest of the house $V_r = \frac{dc_r}{dt} = Q_2c_2 + Q_4c_k - (Q_2 + Q_4)c_r$

This is a set of coupled first order ODEs, which can be re-written as

$$\frac{dc_k}{dt} = \left(\frac{1}{V_k}\right)\cdot\left(Q_1c_o + Q_3c_r - (Q_1 + Q_3)c_k + s\right)$$

$$\frac{dCR}{dt} = \left(\frac{1}{V_r}\right)\cdot\left(Q_2c_o + Q_4c_k - (Q_2 + Q_4)c_r\right)$$

These could be solved by standard DE packages, or one can use elimination them to convert to a single second order DE (See, for Example, E. Kreyszig, *Advanced Engineering Mathematics*, Fourth Edition, Wiley 1979 p132), to get

$$\frac{d^2c_k}{dt} = \frac{dc_k}{dt}\{\gamma + \eta\} + c_k(\beta\varepsilon - \gamma\eta) - \eta\alpha$$

where $\alpha, \beta, \gamma, \varepsilon$, and η are combinations of the constants in the above equations, which could be solved analytically. The simplest procedure is to solve the two first order equations on a spreadsheet by Euler's first method, finding steady state concentrations of 1095 and 537 µg/m^3, which when converted to 24 hour average are practically the same as those found in the previous example. The intermediate values are found as well.

15.16 The house volume is 1500·8 = 12,000 ft^3.

$$Q = 90{,}000\frac{\text{Btu}}{\text{hr}}\cdot\frac{\text{lb methane}}{21{,}502\,\text{Btu}}\cdot\frac{\text{lb mol methane}}{16\text{ lb methane}} = \frac{0.26\,\text{lb mol methane}}{\text{hr}}$$

$$\text{Air Flow} = 11.43\frac{\text{mols air}}{\text{mol methane}}\cdot\frac{0.26\,\text{lb mol methane}}{\text{hr}} = 3.02\frac{\text{lb mol air}}{\text{hr}}$$

We estimate the amount of dilution air by observing that its temperature changes from 70 to 250°F while that of the combustion gas changes from 750 to 250 °F, so that

$$\text{Dilution air} = \text{Combustion air}\cdot\left(\frac{750 - 250}{250 - 70}\right) = 3.02\cdot\left(\frac{500}{180}\right) = 8.4\frac{\text{lb mol}}{\text{hr}}$$

Thus the total air flow is

$$Q = (3.02 + 8.4)\frac{\text{lb mol}}{\text{hr}}\cdot\frac{\text{ft}^3}{2.59\text{ E-3 lb mol}} = 4.4\text{ E3}\frac{\text{ft}^3}{\text{hr}} \approx \frac{1}{3}V_{\text{house}}$$

When the furnace is running it uses about 1/3 of a house volume per hour. The fraction of the time that the furnace is running depends on the outdoor temperature and the insulating properties of the house. Furnaces seldom run more than 50% of the time, even in the coldest weather. This means that for most houses, infiltration air is enough to supply the air to the furnace. For very tight houses a separate air supply for the furnace is required.

The heating code has strong requirements on the supply of air to furnace rooms, to avoid causing a high enough vacuum there to pull exhaust gas into the dwelling.

15.17** $\text{Cost} = \dot{m}C_p\Delta T \cdot \text{Fuel Price} = Q\rho C_p\Delta T \cdot \text{FP}$

$$\text{Cost} = \frac{(1500 \cdot 8)\text{ft}^3}{\text{hr}} \cdot \left(0.075\frac{\text{lb}}{\text{ft}^3}\right) \cdot \left(\frac{0.25\text{ Btu}}{\text{lb} \cdot {}^\circ\text{F}}\right) \cdot (40^\circ\text{F}) \cdot \left(\frac{\$7}{1\text{ E6 Btu}}\right) = \frac{\$0.63}{\text{hr}}$$

$$= \frac{\$1.51}{\text{day}} = \frac{\$45}{\text{month}}$$

15.18** (a) $\text{pCi} = 3.7\text{ E10}\frac{\text{d}}{\text{s}} \cdot 1\text{ E-12} = 0.037\frac{\text{d}}{\text{s}}$

$$\frac{dc}{dt} = -kc; \quad k = \frac{\ln 2}{t_{\frac{1}{2}}}; \quad \frac{dc}{dt} = -\left(\frac{\ln 2}{t_{\frac{1}{2}}}\right) \cdot c; \quad \frac{-0.037}{\text{s}} = \frac{-\ln 2}{3.8\text{ days}} \cdot c$$

$$c = \frac{3.8 \cdot 24 \cdot 3600\text{ s}}{\ln 2} \cdot \frac{0.037}{\text{s}} = 1.75\text{ E4 atoms}$$

(b)

$$y_{\text{radon}} = \frac{4 \cdot 1.75\text{ E4}\frac{\text{atoms}}{\text{L}}}{\frac{6.023\text{ E23 atoms}}{24.055\text{L}}} = 2.8\text{ E-18}$$

15.19** (a) Start with Eq 15.2, with $R = 0$.

$$S = Q(c_i - c_o) = \frac{4000\text{ ft}^3}{\text{hr}}(6-1)\frac{\text{pCi}}{\text{L}} \cdot 28.31\frac{\text{L}}{\text{ft}^3} = 5.66\text{ E5}\frac{\text{pCi}}{\text{hr}}$$

$$S = c_{\text{subsoil}} \cdot Q_{\text{infiltration}}$$

$$Q_{\text{infiltration}} = \frac{5.66\text{ E5}\frac{\text{pCi}}{\text{hr}}}{1000\frac{\text{pCi}}{\text{L}}} = 566\frac{\text{L}}{\text{hr}} = 20^3\frac{\text{ft}}{\text{hr}}$$

(b) Assuming that the infiltration rate is unchanged

$$Q = \frac{S}{c_i - c_o} = \frac{5.66\text{ E5}\frac{\text{pCi}}{\text{hr}}}{(4-1)\frac{\text{pCi}}{\text{L}}} = 1.887\text{ E5}\frac{\text{L}}{\text{hr}} = 6.66\text{ E3}\frac{\text{ft}^3}{\text{hr}} = 111\frac{\text{ft}^3}{\text{min}}$$

which is about the flowrate of an ordinary kitchen fan.

(c) See the solution to Prob. 15.18

$$R = Vc\frac{\ln 2}{t_{\frac{1}{2}}}$$

$$= \left(8000\,\text{ft}^3\right)\cdot\left(\frac{6\,\text{pCi}}{\text{L}}\right)\cdot\left(\frac{\ln 2}{3.8\text{ days}}\right)\cdot\frac{\text{day}}{24\text{ hr}}\cdot\frac{28.31\text{L}}{\text{ft}^3} = 1.03\text{ E}4\frac{\text{pCi}}{\text{hr}}$$

This is 1.03/56.5 = 0.018 2% of S, which I consider negligible

(d)Seal the basement better against infiltration, or pressurize the basement so that leakage is out rather than in, or drive perforated ventilation pipes under the house and pull from them thus lowering the radon concentration and/or pressure there.

15.20 See the solution to Problem 15.19 (c)

$$\frac{R}{Q} = \frac{hAc\frac{\ln 2}{t_{\frac{1}{2}}}}{\frac{hA}{\Delta t}} = \frac{c\ln 2\,\Delta t}{t_{\frac{1}{2}}} = 2.24\frac{\text{pCi}}{\text{L}}\cdot\frac{\ln 2}{3.8\text{ days}}\cdot 1\,\text{hr}\cdot\frac{\text{day}}{24\text{ hr}} = 0.017\frac{\text{pCi}}{\text{L}}$$

$$= \frac{0.017}{2.24} = 0.0076 = 0.76\%$$

15.21 (a)

$$S = 3000\frac{\text{kcal}}{\text{hr}}\cdot\frac{\text{lb}}{21{,}502\,\text{Btu}}\cdot\frac{\text{lb mol}}{16\text{ lb}}\cdot\frac{385\,\text{ft}^3}{\text{lb mol}}\cdot\frac{\text{Btu}}{0.252\,\text{kcal}}\cdot\frac{28.31\text{L}}{\text{ft}^3}\cdot 23\frac{\text{pCi}}{\text{L}}$$

$$= 8671\frac{\text{pCi}}{\text{hr}}$$

$$c_i = \frac{1\text{pCi}}{\text{L}} + \frac{8671\frac{\text{pCi}}{\text{hr}}}{3000\frac{\text{ft}^3}{\text{hr}}}\cdot\frac{\text{ft}^3}{28.31\text{ L}} = 1 + 0.1 = 1.1\frac{\text{pCi}}{\text{L}}$$

(b) Compared to other sources or to the action level, this seems negligible.

Appendix E

When I began teaching this course the students were almost all Chemical Engineers. They had little trouble with the material in this appendix, and found it interesting. As more and more students from other disciplines came into the course I found it harder to use this material because they lacked the background to deal with it. You should feel free to assign this material to Chemical Engineering seniors with little class preparation, but probably should not assign it to others without additional background. I moved it from the main text to an appendix so that the main text would be accessible to students from disciplines other than Chemical Engineering.

E.1 The changes are:

Y becomes $C_{P\,Gas} \cdot T_{Gas}$; w becomes $C_{P\,solid} \cdot T_{solid}$

r becomes Q; k_G becomes h; p_A becomes T_{Gas}

p_A^* becomes $T_{solid\ equilibrium}$

so that the revised equations are

$$\text{E.1; } GA\,C_{P\,Gas}\left(T_{G\,in} - T_{G\,out}\right)\Delta t = A\Delta x\,\Delta\left(\rho_B C_{P\,Sold} T_{Solid} + \rho_G F C_{P\,Gas} T_{Gas}\right)$$

$$\text{E.2; } GC_{P\,Gas}\frac{dT_{Gas}}{dx} = \frac{d\left(\rho_B C_{P\,solid} T_{Solid} + \rho_G F C_{P\,Gas} T_{Gas}\right)}{dt}$$

$$\text{E.3; } -GC_{P_G}\frac{dT_G}{dx} = \rho_B C_{P\,solid}\frac{d\,T_{solid}}{dt};\ \text{E.4; } Q = ha_v\left(T_g - T_s\right)$$

E.5; not used

$$\text{E.6; } T_{gas} = T_{solid\ equilibrium}$$

E.2 As k_G becomes infinite, a and b both become infinite. If we let k_G increase by a factor of a thousand, then $N = ax$ becomes 57,600. This is off the right of Fig. 10.14. However we can see that as N increases the curves are approaching vertical lines, so we may assume that the breakthrough curve is vertical. Furthermore we can see that for large values of N each of the **N** = constant curves crosses the horizontal 0.5 line at bt = **N**, so that we may conclude that here bt **N** = 57,600. We also know that b increases by a factor of 100 and becomes = 7,400/h, so that t = 57,600/7,400 = 7.78 hr.

For infinite mass transfer we would have the ideal breakthrough shown in Fig. 10.13, and the answer would be 8 hr. Here 7.78 hr 8 hr.

As k_G becomes zero, a and b both become zero. If we let k_G decrease by a factor of a thousand, then **N** = ax becomes 0.057. That is too small for Fig 10.14, so we let it decrease by a factor of 7.2, making **N** = ax = 57.6 / 7.2 = 8 From the figure we read that this corresponds to bt = 1. Here b = 7.4 / 7.2 1/hr , so that t 1/1 = 1 hr.

From the shape of the curves on Fig. 10.14 we see that as the mass transfer coefficient becomes smaller and smaller the curves depart more and more from the vertical shape they would have for the ideal situation sketched in Fig 10.13, and, all other things being equal, breakthrough would occur sooner and sooner. Then we can see that at the limit as the mass transfer coefficient approaches zero, none of the adsorbate would be adsorbed and breakthrough would occur as soon as the original (presumed inert) material in the bed had been mechanically displaced.

E.3 As closely as I can read Fig. 10.14, the **N** = 60 line crosses the horizontal 0.1 line at bt 45. Thus the breakthrough time is the time in example multiplied by (45/36) = 1.25, or 4.9 x 1.25 = 6.1 hr. This makes sense; if one looks as the "Real" curve on Figure 10.13 one sees that it reaches higher values later than it reaches lower values.

E.4 Here we read at **N** = ax 60/2 = 30, and breakthrough time for which bt = 36, finding $Y_{a\ out} / Y_{a\ in}$ 0.8. Thus, halfway through the bed the solid is not completely saturated yet, and some transfer from the gas to the solid is still occurring.

If we had asked for the concentration 10% of the way through the bed we would read at **N** = ax 60/10 = 6 finding that $Y_{a\ out} / Y_{a\ in} > 0.999$. Thus, at that part of the bed the solid is practically saturated, and the inlet gas passes through it without transferring any toluene to it.

E.5 For breakthrough defined as 1% of the inlet, the concentration of interest is 1% of 5000 ppm = 50 ppm. This is a small enough value that we cannot find w^* from Fig. 10.10. Instead we compute the saturation parameter as in Example 10.11, finding

$$\text{Saturation Parameter} = \frac{(560)\cdot(0.782)}{1.8\cdot 92}\log_{10}\left(\frac{0.07}{50\text{ E - 6}}\right) = 8.32$$

Then we read curve F on Figure 10.11, finding an intercept of 15, from which we calculate

$$w = 15\frac{\rho_L}{100} = 0.15\cdot(0.782) = 0.117: \quad c = \frac{\frac{p_a^*}{P}}{w} = \frac{50\text{ E - 6}}{0.117} = 4.3\text{ E - 4}$$

This value of c is much less than that in Example 10.11, indicating a steeper slope of the line from the intercept to the value on Figure 10.11. Changing c does not change a. But b is proportional to it, so that the revised value of b is 0.185/hr. For bt = 36, we find

$$t = \frac{bt}{b} = \frac{36}{\frac{0.185}{\text{hr}}} = 194.5\text{ hr}$$

This appears odd, but is sensible. Here for this choice of c we can absorb 40 times as much toluene per gram of adsorbent (at the same concentration) as in that example, so we would expect a breakthrough time, based on material balance alone, of 40 x 4.9 = 196 hours. The mass transfer will be perturbed as well, but this shows the effect.

Since the capacity of the bed is a critical parameter, we should always use the value of c corresponding to inlet (which means, given the shape of the curves on Fig. 10.12, the lowest equilibrium loading of the solid).

E.6 (a) The bed cross sectional area becomes 16 ft^2, thus reducing the mass velocity by a factor of 4. The particle Reynolds number is reduced by a factor of 4 so that the Chilton-Colburn analogy factor is multiplied by $(1/4)^{-0.3} = 1.52$. The mass transfer coefficient is multiplied by (1.52/4). a is proportional to (mass transfer coefficient / mass velocity) so its value becomes $a = 14.4\text{ft}\cdot\left(\frac{1}{1/4}\right)\cdot\left(\frac{1.52}{4}\right) = \frac{21.9}{\text{ft}}$

b is proportional both to a and to G so its value becomes $b = \frac{7.4}{\text{hr}}\cdot\frac{1}{4}\cdot 1.52 = \frac{2.81}{\text{hr}}$

N = ax = 21.9/ft x 1 ft = 29.1. Reading a breakthrough of 0.01 for this value of **N** we find bt 8, so that $t = \frac{bt}{b} = \frac{8}{2.81\text{ hr}} = 2.85\text{ hr}$

This is 58% of the time in the example, indicating that with this shallower bed we would expect the imperfect mass transfer to manifest itself at the outlet side sooner.

(b) See problem 10.25. The pressure drop is proportional to the bed length times the superficial velocity. Thus the answer is the answer from that problem (0.5 psi) times (1/4)(1/4) = 0.031 psi.

In the most careful work we would observe that the particle Reynolds number is reduced by a factor of 4 to about 24, which would lower the ratio of the observed pressure drop to that of the Kozeny-Carman equation from about 2 to perhaps 1.8. It is doubtful if we should consider this, because of the many other uncertainties in this calculation.

This problem does show that a broad shallow bed has a much lower pressure drop than a long, narrow one, but that the long, narrow one has a better breakthrough performance. Engineers tailor their bed geometry to the constraints of the problem they face.

End of Solutions Manual.

Again, I will be grateful to those who point out to me errors in the text or in this manual.

I hope you enjoy teaching this course as much as I have in the past, and expect to in the future.

N de N